Preface to the Series

The *Corrosion Testing Made Easy* series was conceived in 1983 by members of the NACE Publications Committee in recognition of the need for easy-to-follow guidelines on specific types of corrosion testing. While many test procedures are defined in NACE and other industry standards and some textbooks, it was noted that none of these sources was written with the beginner in mind; all assumed some prior experience with corrosion testing and familiarity with corrosion terminology.

This situation inevitably leads to a need for the entry-level technician to be trained initially by a more experienced experimentalist. The cost of such training can be quite high, especially if a highly paid professional scientist or engineer must do the training. Furthermore, because of demands on their time, the more experienced experimentalists may not always be available to give timely advice. Consequently, there seemed to be an urgent need for a series of books that could guide an entry-level technician through a corrosion test procedure with little or no assistance. These books would make no assumptions about prior knowledge or expertise. They would describe the possible difficulties in performing the experiment and would indicate how to correct mistakes if they were made. The books would each contain an unusually large number of illustrations on the grounds that "one good picture is worth a thousand words." Hence, the *Corrosion Testing Made Easy* series was born.

Each author for the series was chosen with great care. All authors are technically very competent, all have had many years of hands-on experience in performing corrosion testing, and, perhaps even more important, they have a reputation for clear, concise writing.

Most of the books in the series can be used as step-by-step guides to performing specific types of corrosion tests, e.g., stress corrosion tests or electrochemical tests. However, the introductory volume, *The Basics*, describes techniques and equipment that are common to most, if not all, corrosion tests. It provides a glossary that relates to corrosion, and it contains some basic principles of corrosion and electrochemical theory. Although the practical aspects of corrosion testing are emphasized in all volumes in the series, theoretical aspects are introduced when they can help the experimentalist perform a test with greater ease.

More than a dozen volumes are planned for publication in the next few years. While they are intended primarily for entry-level technicians, it is clear that these books may also be useful for more experienced scientists and engineers. For instance, engineers skilled in the art of performing fatigue tests may use the *Corrosion Testing Made Easy* series to obtain practical advice on extending their knowledge of corrosion fatigue tests. Or an experienced biologist may need help in performing a seawater corrosion test as part of a larger study on marine toxicology. Whoever uses the series, one thing is certain: The volumes are unique and fill an important gap in the scientific literature.

Barry C. Syrett

Corrosion Testing Made Easy — Volume 2

Galvanic Corrosion Test Methods

Harvey P. Hack

Naval Surface Warfare Center
Annapolis, Maryland

To Aunt Gladys + Uncle Ed,
The artwork isn't up to your standards, but I hope you enjoy this anyway.

Harvey

Published by
NACE International
1440 South Creek Drive
Houston, Texas 77084

Carolyn Donahoo
Director, Publications

Aunee Pierce
Senior Editor, Books

Claire Blondeau
Editor, Books

Library of Congress Catalog Card Number: 93-084558
ISBN 1-877914-55-X

To my wife, Laura, and my children, Sarah, James, and Rachel

Preface

There are few standardized galvanic corrosion test procedures because of the large number of factors that can influence this type of corrosion and the many different aspects of galvanic corrosion that can be studied. Therefore, it is important for the technician conducting galvanic corrosion tests to be sufficiently familiar with the theory of galvanic corrosion to understand what can influence test results. This book is not a cookbook of testing procedures, but a set of guidelines concerning the important issues in each type of galvanic corrosion test.

The author is grateful to the editor of the Corrosion Testing Made Easy series, Barry Syrett, and to John Beavers and Laura Hack for their extensive review and comments for improving the clarity of the manuscript.

Harvey P. Hack
November 1993

Contents

CAUTION

The test methods described in this publication may involve hazardous materials, operations, and equipment. This publication does not purport to address all of the safety problems associated with the test methods. It is the responsibility of whoever uses these test methods to consult and establish appropriate safety and health practices and to determine the applicability of regulatory limitations before use.

CHAPTER 1

Introduction

The Nature of Galvanic Corrosion

When a structure made of more than one conductive material is operated in a corrosive environment, an increase in corrosion of some of the materials may occur because of galvanic corrosion.

Galvanic corrosion is a major cause of failure in a wide variety of devices and environments. Galvanic corrosion in the atmosphere leads to corrosion failures in roofing, gutters, and other architectural components. Road salts cause galvanic corrosion of automobile bodies near trim and bumper materials. Seawater splash and spray can lead to galvanic corrosion of ship fittings and deckhouse structures. Seawater immersion can cause failures of ship hull plating, boat propellers and shafts, valves, condensers, and other piping components. Even drinking water can lead to galvanic corrosion of water heaters and dissimilar metal plumbing fixtures. Galvanic corrosion is a major cause of corrosion failure in bolts, rivets, welds, and plated components. In fact, anywhere two or more different metals are used in a structure the potential for galvanic corrosion exists.

Some definitions that professional societies have adopted for galvanic corrosion are:

American Society for Testing and Materials (ASTM) — accelerated corrosion of a metal because of an electrical contact with a more noble metal or nonmetallic conductor in a corrosive environment.[1]

NACE International (formerly National Association of Corrosion Engineers) — corrosion associated with the current resulting from an electrical coupling of dissimilar electrodes in an electrolyte.[2]

Three conditions must be met for metal degradation to be called galvanic corrosion: (1) at least two different conductive materials must be involved, (2) the materials must be in electrical contact so that a current can flow between them, and (3) a corrosive medium, or electrolyte, must be present. If any one of these three conditions is removed, galvanic corrosion will not occur. When these three conditions are met, corrosion of one or more materials may increase while corrosion of the others will decrease or remain unchanged. The increase in corrosion above the rate normally experienced by that material by itself is called galvanic corrosion. The three conditions necessary for galvanic corrosion to occur are shown schematically in Figure 1.1 (see Page 2).

Galvanic corrosion can have beneficial effects. The reduction in corrosion rate of one of the materials in the couple is the basis for sacrificial cathodic protection, which is used to minimize corrosion of tanks, ship hulls, buried or underwater pipelines, and even electrical cables. The current produced by batteries is a result of galvanic corrosion.

Galvanic action can cause, intensify, or inhibit any of the other forms of corrosion. Environmentally induced cracking and corrosion fatigue are examples of corrosion forms whose rate can be changed by galvanic action. Because galvanic corrosion cannot usually be identified by appearance alone, its presence must be inferred from the types of materials involved, the geometrical and electrical arrangement of these materials, and the location and severity of the corrosion compared to that expected if only one material were involved. Fortunately, once galvanic corrosion has been identified as the cause of a problem, the ways of minimizing the attack are well known.

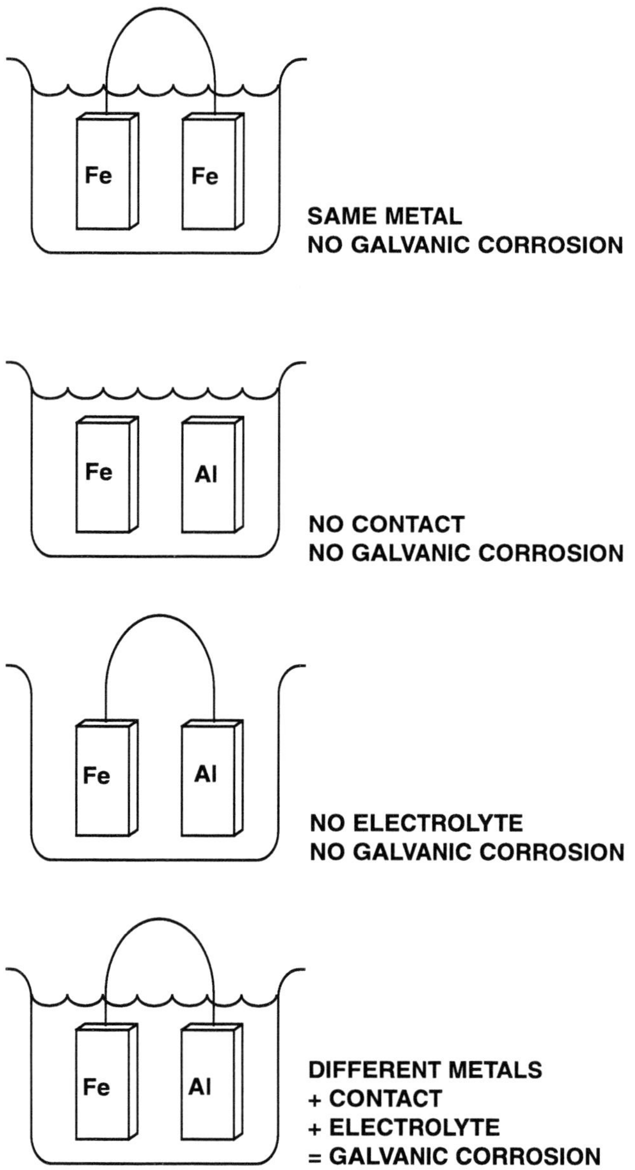

FIGURE 1.1 - Conditions necessary for galvanic corrosion.

Galvanic Corrosion Testing

The amount of other forms of corrosion can be changed by galvanic action. Any of the tests used for corrosion — such as uniform attack, pitting, or stress corrosion described in other volumes of the *Corrosion Testing Made Easy* series — can be used, with modifications, to determine galvanic corrosion effects. The modification is frequently as simple as connecting a second metal to the one in test with a wire, or it can be as complex as the use of sophisticated electrochemical apparatus. Sometimes, all that is needed to convert a conventional corrosion test into a galvanic corrosion test is a change in the method of data interpretation.

Galvanic corrosion testing can be performed on full-scale components, reduced-size models, laboratory coupons, or electrochemical specimens. In the next chapter, certain features common to all these types of tests will be discussed, and the remaining chapters will describe each of these test types in detail.

Galvanic Corrosion Theory

The Importance of Theory

Many standardized corrosion tests have been specified step-by-step in written standards, such as those of NACE and ASTM. Standard tests for stress corrosion include the C-ring test, modified wedge-opening loaded test, and bent beam test. Standard crevice corrosion tests include the ferric-chloride test and the multiple crevice assembly. A standard test for exfoliation corrosion is the Asset test; for intergranular corrosion, the Streicher test. Atmospheric testing includes the use of a number of standardized salt spray and other cabinet tests.

Galvanic corrosion has few standard tests, however. Because there are so many variables to consider in galvanic corrosion testing, standardization is difficult. Also, there are many ways to predict galvanic corrosion, depending on which test variables can be controlled and which can be measured. Galvanic corrosion testing may have a different test design for each specific situation being modeled. Therefore, test design should be carried out by someone well versed in the theory of galvanic corrosion. This chapter will present sufficient theory to allow the reader to understand the reasons for and important factors in galvanic corrosion test design alternatives. Only in Chapter 5 will detailed step-by-step standardized testing be discussed.

Electricity and Corrosion

When a single piece of metal corrodes, its atoms go into solution in the surrounding environment called the electrolyte. The driving force for this phenomenon is a lowering of the free energy of the metal/electrolyte system. An atom of the metal has no net charge, while a metal ion in solution in an electrolyte such as water will usually have a positive charge. Therefore, during corrosion an atom must gain positive charge by the loss of at least one electron (Figure 1.2). This type of reaction is called oxidation even if oxygen takes no part in the reaction. The electrons produced by the oxidation reaction must be consumed by a reduction reaction, so called because the charge on the ions is reduced. Because electrons are transferred during corrosion reactions, these reactions are called electrochemical reactions. Two common reduction reactions that use up electrons generated during corrosion in water are (1) oxygen reduction—oxygen at zero charge dissolved in the electrolyte combines with electrons and water to create negative hydroxyl ions; and (2) hydrogen reduction—typically in acids, the positively charged hydrogen ions in the solution combine with electrons and each other to form hydrogen gas (Figure 1.3).

Corrosion thus generates a flow of electrical current within the metal between sites where oxidation occurs and sites where reduction occurs. For corrosion to proceed, the reduction reactions must use up electrons at the same rate the oxidation reactions create them. The balance between these reactions results in the metal exhibiting an electrochemical potential. The value of this potential is uniquely set for each metal/electrolyte system and, in the absence of an externally applied current, is called the corrosion potential, abbreviated E_{corr}.

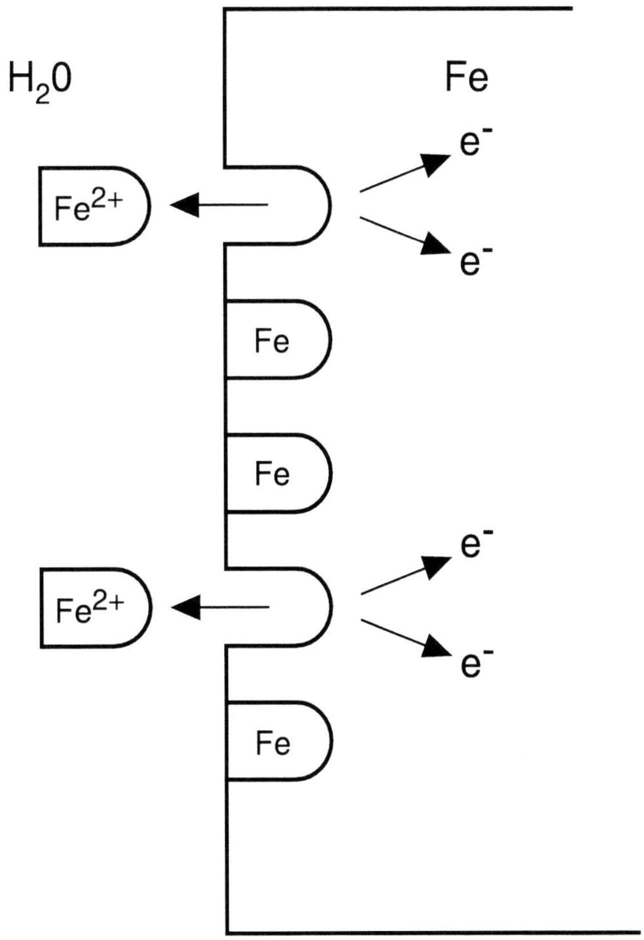

FIGURE 1.2 - Iron oxidation.

Oxygen Reduction Reaction:

$$O_2 + 4e^- + 2H_2O \longrightarrow 4OH^- \qquad \text{4 electrons consumed}$$

(O_2: zero charge on each O; $4OH^-$: -2 charge on each O)

Hydrogen Reduction Reaction:

$$2H^+ + 2e^- \longrightarrow H_2 \text{ (gas)} \qquad \text{2 electrons consumed}$$

($2H^+$: +1 charge on each H; H_2: zero charge on each H)

FIGURE 1.3 - Common reduction reactions in water.

Determining the Corrosion Rate

Current vs Corrosion—Faraday's Law

Because electrons are created by oxidation and used up by reduction reactions, an electron current must exist between the sites of the two types of reactions. Under undisturbed, or free corrosion, conditions, this electron flow results in a conventional current, called the corrosion current, in the opposite direction (Figure 1.4). To avoid accumulation of charge, the corrosion current is balanced by an equal ionic current in the electrolyte. The corrosion current is frequently divided by the surface area and expressed as a corrosion current density. The corrosion current density is a measure of how rapidly oxidation, and therefore loss of metal into solution, is occurring under free corrosion conditions. The corrosion current can be used to calculate the amount of mass loss due to corrosion by applying Faraday's Law:

$$W = I\,t\,E$$

where: W = mass loss due to corrosion (g)

I = corrosion current (A)

t = time current has been flowing (s)

E = electrochemical equivalent: the mass of metal ionized by one ampere flowing for one second (g/A-s). This is a function of the density, atomic weight, and change in charge during oxidation, and is calculated as:

$$E = \frac{m}{nF}$$

where: m = atomic weight (g/mo)

n = change in the charge of the metallic species (eq/mo)

F = Faraday's Constant (96,501 A-s/eq)

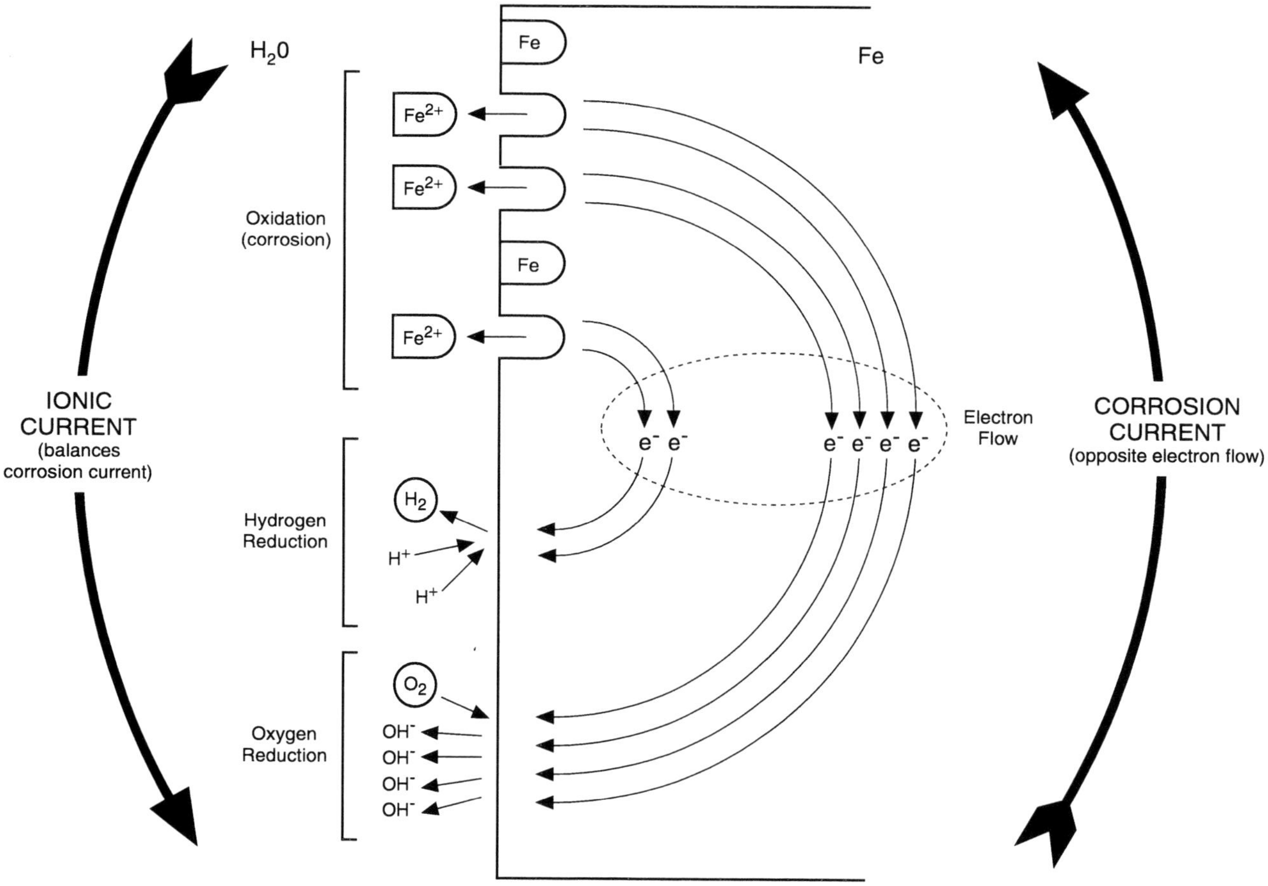

FIGURE 1.4 - Oxidation and reduction during corrosion of iron.

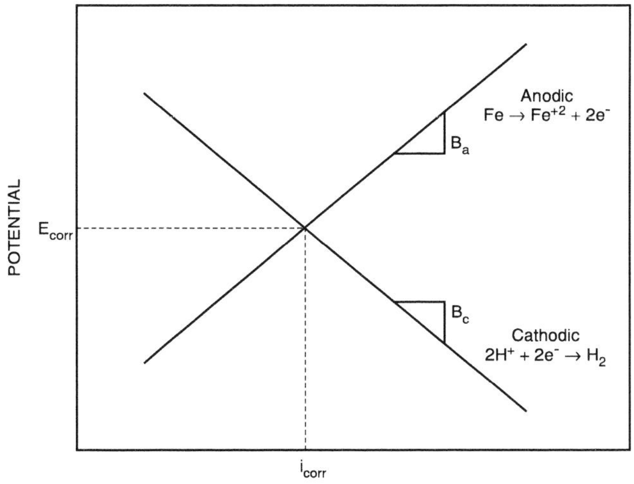

FIGURE 1.5 - Evans diagram for iron in acid.

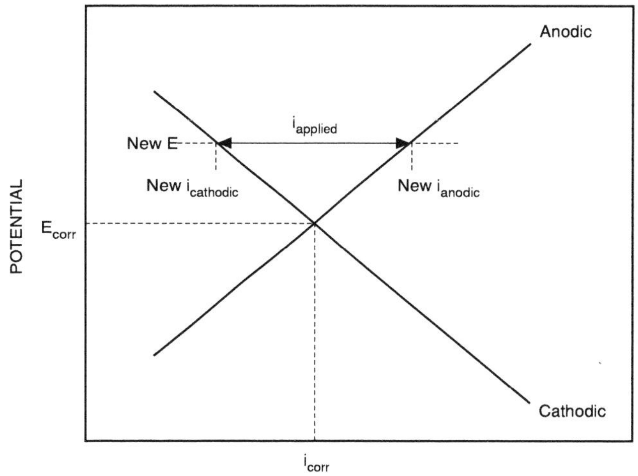

FIGURE 1.6 - Evans diagram showing change of anodic rate caused by an applied current.

Corrosion rate is mass loss per unit time, which can be related to corrosion current density for metals under free corrosion conditions. It is unfortunate that the corrosion current flows between anodic and cathodic sites entirely within the corroding metal—not through an external wire where it could be directly measured. Indirect measurements are possible, however, and this has led to the development of various types of corrosion rate meters.

Current vs Potential

Corrosion usually slows down when the electrochemical potential of the metal is made more negative by applying a current from an outside source. Tafel observed that the rate of the corrosion reaction expressed as a current density is frequently a logarithmic function of potential. Tafel's Law is expressed as:

$$E-E_{corr} = B \log i/i_{corr}$$

where: E_{corr} = corrosion potential

i_{corr} = corrosion current density

i = current density at potential E

E = potential of the metal

Figure 1.5 is a plot of potential vs log of current density. On such a plot, called an Evans Diagram, Tafel's Law predicts a straight line with a slope of B, called the Tafel Slope. All oxidation, or anodic, reactions that follow Tafel's Law will have a positive slope on such a diagram. All reduction, or cathodic, reactions will have a negative slope on the same diagram. Both anodic and cathodic reactions occur on any metal surface under free corrosion conditions at an equal rate, which is at the intersection of the two curves, the corrosion current.

If the metal's potential is shifted by applying a current from an outside source, the anodic and cathodic reaction currents will shift in opposite directions (Figure 1.6). To maintain charge neutrality, the applied current must be equal to the difference between the anodic current and the cathodic current:

$$i_{applied} = i_{anodic} - i_{cathodic}$$

At potentials very positive to the corrosion potential, the cathodic current will be very small relative to the anodic current, making the applied current almost equal to the anodic current. At potentials very negative to the corrosion potential, the anodic current will be very small relative to the cathodic current, making the applied current almost equal to the cathodic current. At the corrosion potential, the anodic and cathodic currents are equal, making the applied current zero. A plot of applied current density vs potential, as shown in Figure 1.7, is called a polarization diagram or polarization curve. Polarization diagrams are particularly useful because they relate an applied current that can be measured to a measured potential.

An equation for an ideal polarization diagram in which anodic and cathodic reactions both follow Tafel's Law is plotted in Figure 1.7 and is called the Butler-Volmer Equation:

$$i_{applied} = I_{corr} [\exp\{(E-E_{corr})/B_a\}-\exp\{-(E-E_{corr})/B_c\}]$$

where: B_a = Anodic Tafel Slope

B_c = Cathodic Tafel Slope

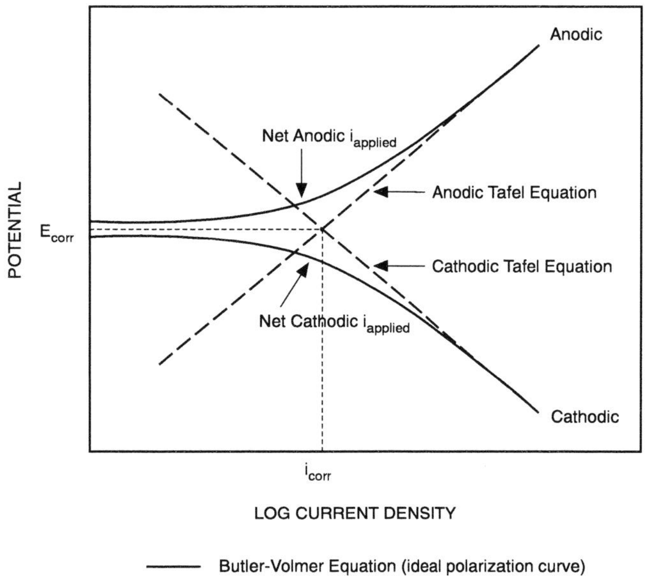

FIGURE 1.7 - Evans diagram showing relationship between observed polarization curve and Tafel equations.

The Galvanic Couple

So far only one metal surface has been looked at. When two different metals are in an electrolyte, each one has its own anodic and cathodic reactions, crossing at a specific corrosion potential at a specific corrosion current density. This is illustrated in Figure 1.8 for iron and copper in water. If the two metals are electrically connected, for example, by a copper wire or direct contact, and if their corrosion potentials are different, a current will flow through the electrical path from the more positive metal to the more negative metal to try to equalize the potentials. The total current flowing out of one metal must equal the total current flowing into the other, even if the surface areas of the two metals are not equal.

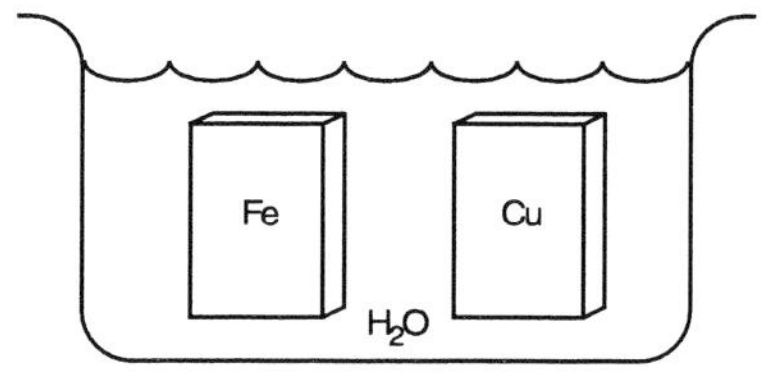

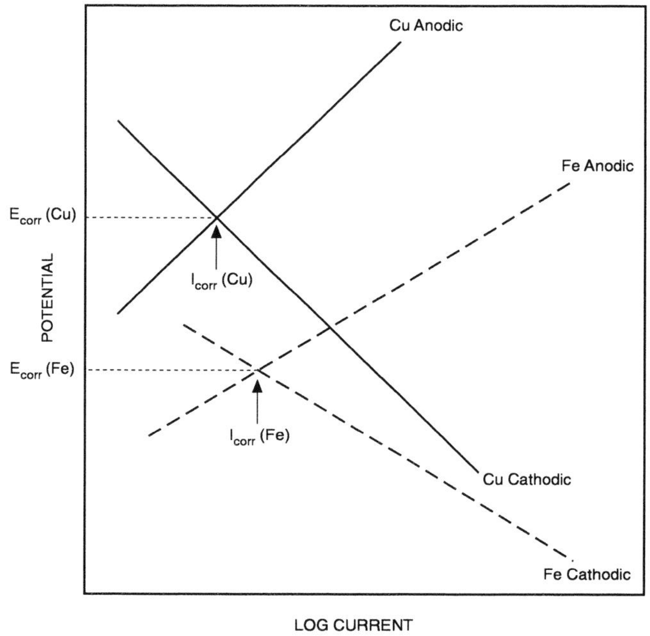

FIGURE 1.8 - Evans diagram for two uncoupled metals.

For that reason, total current rather than current density is plotted in Figure 1.8 and whenever different metals are considered on the same plot. The sum of the anodic currents on both pieces of metal must equal the sum of the cathodic currents to maintain charge neutrality. The anodic sum and the cathodic sum can be plotted on an Evans Diagram, as in Figure 1.9. Their intersection point defines the common potential of the coupled metals—called the coupled potential or mixed potential of the couple—and the current that flows between all anode and cathode reactions, called the total current. The potential of each metal has been shifted from its E_{corr} to a new value, E_{couple}. The corrosion rate of each metal has also been changed because of the couple, from I_{corr} to a new value, I_{couple}, which can be picked off the anodic curve for that metal at E_{couple}. The rate on the initially more negative material (iron in Figure 1.9) increases, whereas the rate on the initially more positive material (copper in Figure 1.9) decreases. The

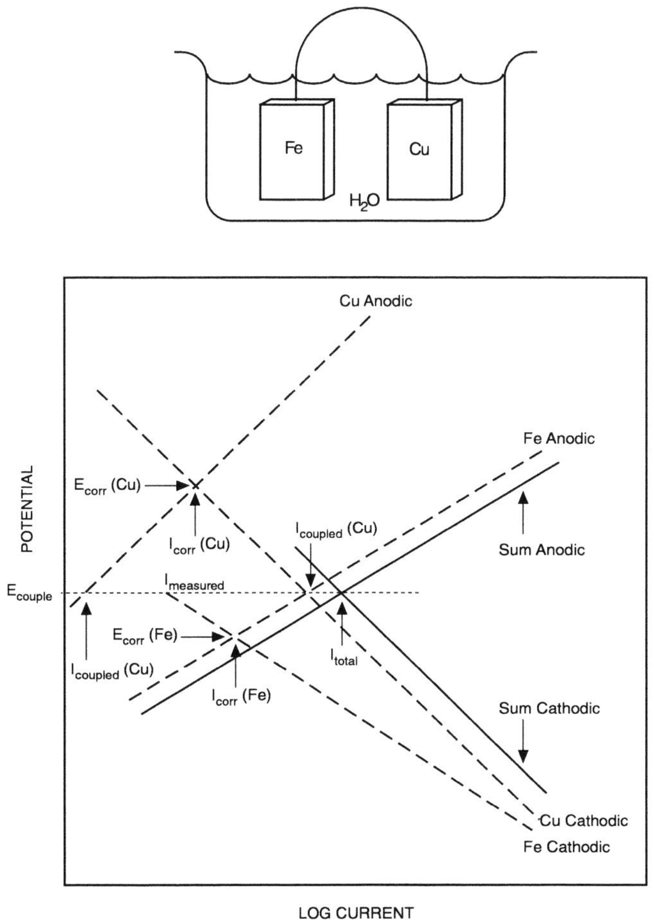

FIGURE 1.9 - Evans diagram for two coupled metals.

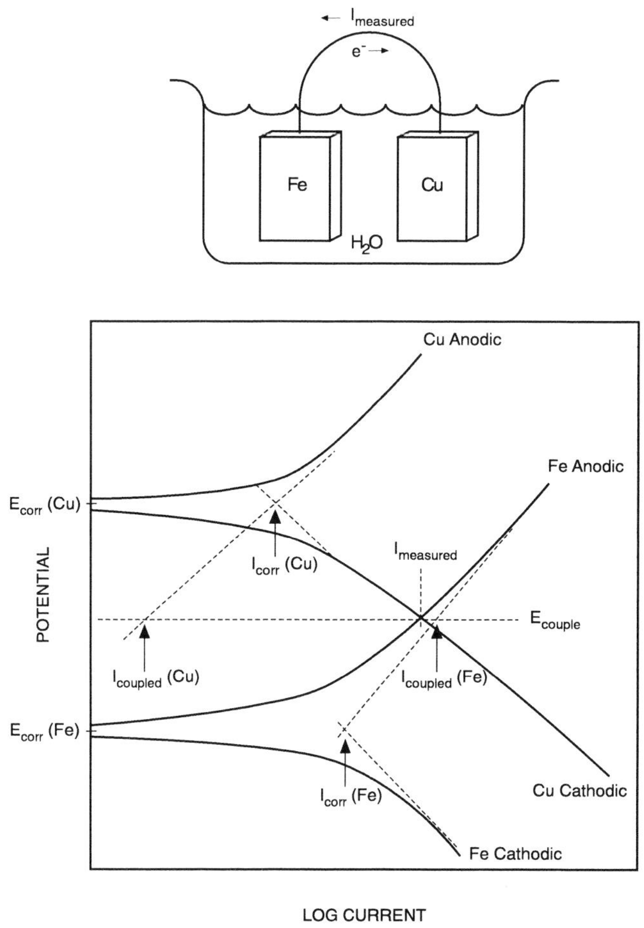

FIGURE 1.10 - Polarization for two coupled metals.

increase in corrosion rate above the freely corroding rate is called galvanic corrosion. The decrease in corrosion rate of the other metal is called cathodic protection and is an extremely useful and widespread method of controlling corrosion.

It is difficult to visualize the current that flows through the wire connecting the two metals using Evans diagrams. To view this current better, a polarization diagram can be used. In Figure 1.10 polarization curves for each of two metals are plotted together. When the metals are electrically connected, current flowing from the more cathodic material must be equal to the current flowing into the more anodic material. This current is the galvanic current, $I_{galvanic}$. This occurs at the intersection of the anodic branch of the more negative material (the anode) with the cathodic branch of the more positive material (the cathode). This intersection defines the coupled potential and the galvanic current.

Polarization diagrams can better show the galvanic current because they are more closely related to measurable currents than Evans diagrams, but they are not as good at showing changes in corrosion rate. Galvanic current, $I_{galvanic}$, is frequently incorrectly used to predict galvanic corrosion rate. As can be seen in Figure 1.10, the true corrosion rate is I_{couple}. The true rate may be close to $I_{galvanic}$, but this is not always the case.

Complicating Factors

The shape of reaction curves need not be a straight Tafel Line on an Evans Diagram as in Figures 1.5 to 1.7. The rate-controlling step in the reaction may change, leading to a change in slope. Frequently, at potentials far from E_{corr}, the reaction rate is limited by movement of reactants through the electrolyte, creating a potential-independent rate. This can be seen as a vertical line on the Evans Diagram. Anodic reactions of some materi-

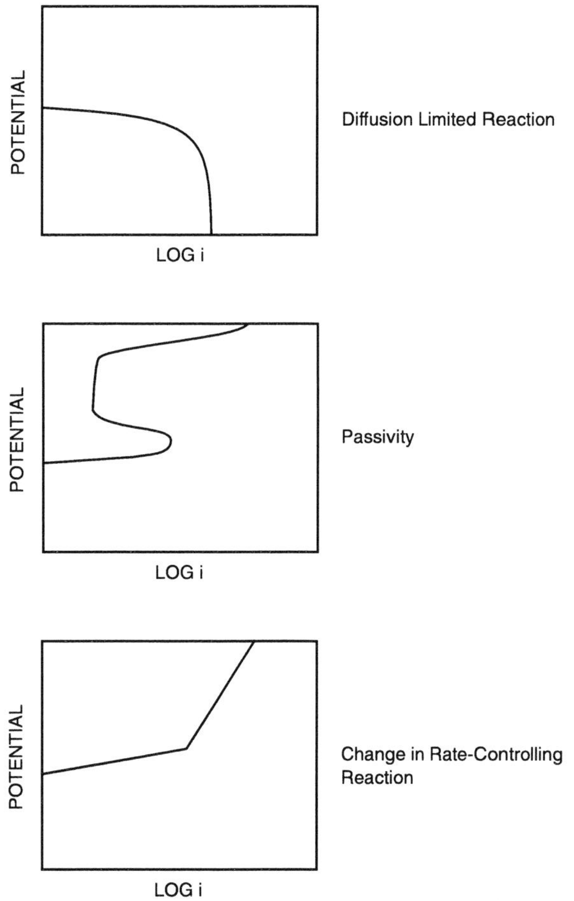

FIGURE 1.11 - Evans diagram of other polarization curve shapes.

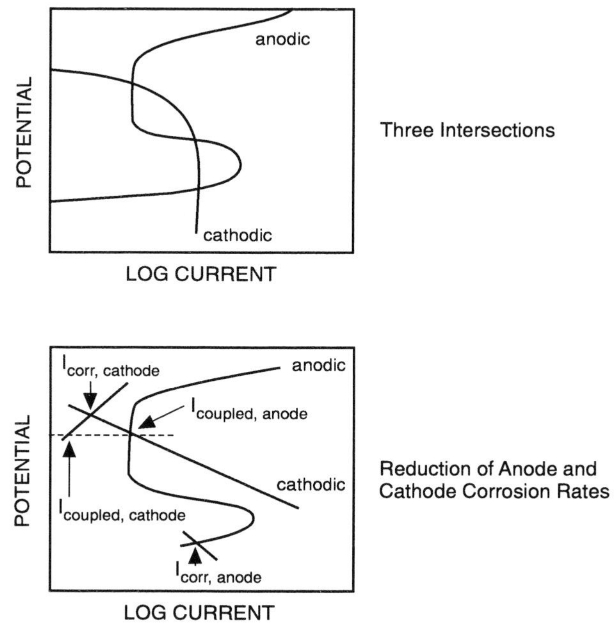

FIGURE 1.12 - Evans diagram for complex galvanic interactions.

als may show a reduction in rate as potential is increased, a phenomenon called passivity. This is caused by a potential being reached where the products of the reaction can accumulate on the metal surface as a tightly adhering, nonporous layer. This creates a barrier to further reaction. These different curve shapes, illustrated in Figure 1.11, can lead to complex galvanic effects such as the possibility of the anode and cathode curves intersecting at three points, or the reduction in corrosion of both anode and cathode upon coupling.

These less-straightforward situations are shown in Figure 1.12. The upper part of the figure shows three intersections of the anode and cathode curves instead of the single intersection that is more common. In this case the galvanic couple can exist in several different states, rapidly switching between the various states with little stimulus. The lower part of the figure shows both the anode and the cathode rate reduced as a result of the couple, because the anode is forced into an area of passivity.

Reaction rate curves such as those in Figures 1.5 to 1.7 are usually plotted as current density rather than total current. When the anode and cathode materials have different surface areas, the curves must be scaled before they are superimposed to get total currents. This is done by multiplying current density by the wetted surface area of that metal. Total currents must be used because it is the total anodic and cathodic currents that must be equal, not the current densities. This is the origin of the area ratio effect, sometimes called the catchment principle. As illustrated in Figure 1.13, a large cathode-to-anode area ratio will result in a higher galvanic corrosion rate than a small ratio. This is why fasteners that are anodic to the material they are holding will corrode so quickly.

When a structure contains more than two materials in electrical contact, polarization curves can still be used to predict galvanic corrosion. The sum of all the anodic reaction currents must still be equal to the sum of all the cathodic reaction currents. The process of predicting galvanic corrosion rates for three dissimilar metals is just an extension of that for two metals.

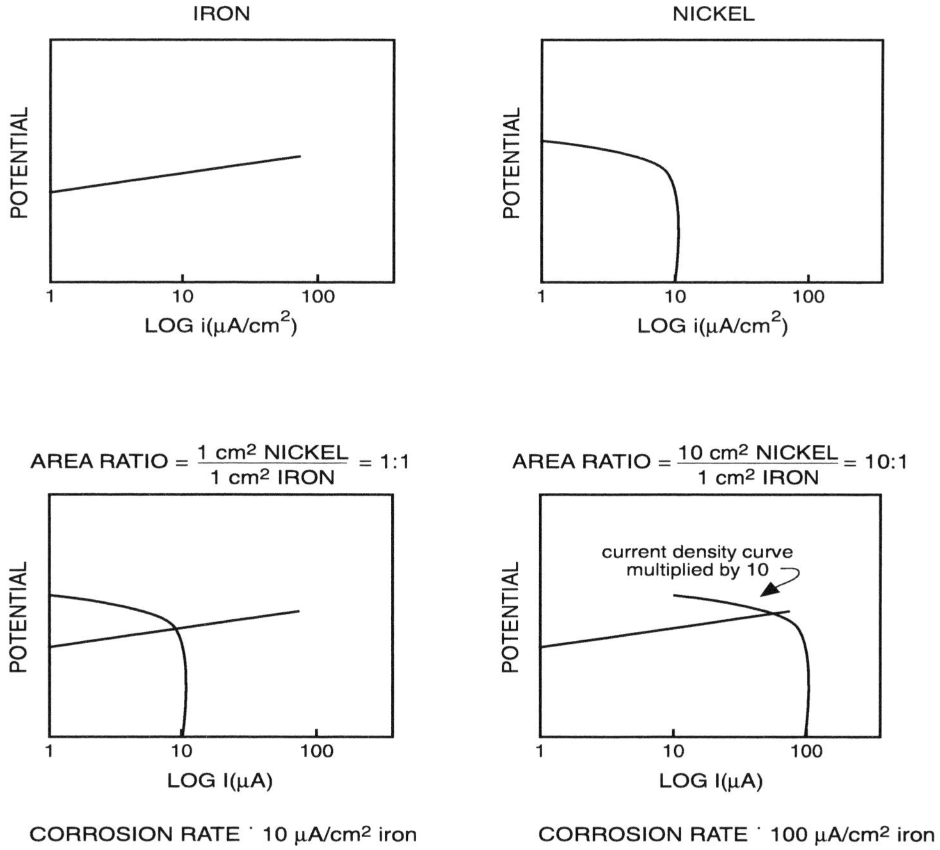

FIGURE 1.13 - Evans diagrams showing rate effect.

Another complicating factor is low conductivity of the electrolyte. If the anode and cathode materials are far enough apart that the electrical resistance of the electrolyte becomes significant, this resistance will slow down galvanic corrosion. Current flowing through a resistive electrolyte will generate a voltage loss, which can be calculated from Ohm's Law: E=IR. This voltage loss is appropriately called IR drop and results in the anode and cathode being at different potentials, separated by that amount. As seen in Figure 1.14, this lowers the rate of the reactions. IR drop is very important to consider when designing galvanic corrosion tests because it can significantly affect the results.

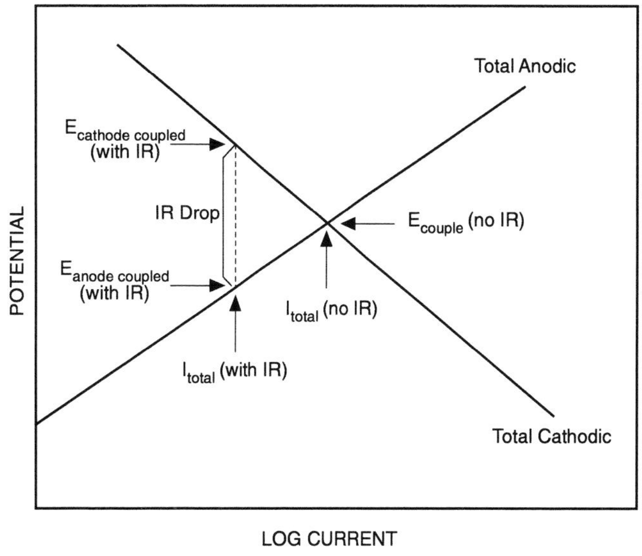

FIGURE 1.14 - Evans diagrams showing effects of IR drop.

References

1. ASTM G15-92. "Standard Terminology Related to Corrosion and Corrosion Testing" (Philadelphia, PA: ASTM, 1992).

2. *NACE Publications Style Manual* 2nd ed. (Houston, TX: NACE, 1989), p. 53.

Bibliography

General

Fontana, M.G. *Corrosion Engineering*. New York, NY: McGraw-Hill, 1986.

Morgan, J. *Cathodic Protection*. 2nd ed. Houston, TX: NACE, 1987.

Rowe, L.C., R.L. Chance. "Fundamentals of Corrosion Testing." *Automotive Corrosion by Deicing Salts*. Ed. R. Baboian. Houston, TX: NACE, 1981.

Shoesmith, D.W. "Kinetics of Aqueous Corrosion." *ASM Metals Handbook*, Vol. 13, 9th ed. Materials Park, Ohio: ASM International, 1987.

Uhlig, H.H. *Corrosion and Corrosion Control*. New York, NY: John Wiley and Sons, Inc., 1971.

Mixed Potential Theory

Baboian, R. "Electrochemical Techniques for Predicting Galvanic Corrosion." *Galvanic and Pitting Corrosion—Field and Laboratory Studies*. R. Baboian, W. France Jr., L. Rowe, and J. Rynewicz, eds. STP 576. Philadelphia, PA: ASTM, 1976.

Baboian, R. "New Methods for Controlling Galvanic Corrosion." *Machine Design*, Oct. 11, 1979: pp. 2-9.

Tomashov, N.D. *Theory of Corrosion and Protection of Metals*. New York, NY: Macmillan Company, 1966.

Passivity

Baboian, R. "Investigations of Galvanically Induced Localized Corrosion." *Localized Corrosion—Cause of Metal Failure*. STP 516. Philadelphia, PA: ASTM, 1972.

Baboian, R., G.S. Haynes. "Galvanic Corrosion of Ferritic Stainless Steels in Seawater." *Corrosion in Natural Environments*. STP 558. Philadelphia, PA: ASTM, 1974.

Area Ratio Effect

Mansfield, F., J. Kenkel. "Laboratory Studies of Galvanic Corrosion of Aluminum Alloys." *Galvanic and Pitting Corrosion—Field and Laboratory Studies*. R. Baboian, W. France Jr., L. Rowe, and J. Rynewicz, eds. STP 576. Philadelphia, PA: ASTM, 1976.

IR Drop and Current Distribution

Compton, K.G. "Current Density Distribution on Composite Structures Under Cathodic Protection in Seawater." *Galvanic and Pitting Corrosion—Field and Laboratory Studies*. R. Baboian, W. France Jr., L. Rowe, and J. Rynewicz, eds. STP 576. Philadelphia, PA: ASTM, 1976.

Hack, H.P., P.J. Moran, J.R. Scully. "Influence of Electrolyte Resistance on Electrochemical Measurements and Procedures to Minimize or Compensate for Resistance Errors." *The Measurement and Correction of Electrolyte Resistance in Electrochemical Tests*. L.L. Scribner and S.R. Taylor, eds. STP 1056. Philadelphia, PA: ASTM.

CHAPTER 2

Factors to Consider When Testing

Making the Test Resemble Reality

Galvanic corrosion can be studied using a variety of tests ranging from full-scale component exposures to electrochemical tests. Testing must take into account other forms of corrosion that could affect the results, for example, pitting or stress corrosion cracking. Nevertheless, there are some common factors to consider in all types of galvanic corrosion tests. This chapter will discuss these factors and their importance in determining the accuracy of the test results. The test conditions must closely simulate those expected in service, so factors that must be considered are materials, environment, and geometry, as shown in Figure 2.1.

MATERIALS
+
ENVIRONMENT
+
GEOMETRY
=
GALVANIC CORROSION TEST

FIGURE 2.1 - Factors to consider in all types of galvanic corrosion tests.

Materials

When testing the strength of materials for use in cars, it is clearly more appropriate to test automotive steels than aircraft steels, as the strength of the two classes of steel can be markedly different. Similarly, in galvanic corrosion testing, the most desirable materials to use in the test are those that are used in service. Even small differences between the materials of concern and those actually tested may have strong effects on the test results. Figure 2.2 shows the three most important material parameters: chemical composition, thermal history, and surface condition (see Page 12). Changes in these parameters may change the corrosion behavior of a material when it is not in a galvanic couple without affecting its galvanic corrosion behavior; or such changes may affect a material's galvanic corrosion behavior without significantly affecting its corrosion properties when it is used alone.

Certain alloying elements can be very potent, for example, copper in weathering steels, in that a small addition can cause a large decrease in corrosion rate in the atmosphere. Small compositional changes that affect the metallurgical structure of the alloy may also lead to profound changes in corrosion behavior. This is evident in the relatively poor corrosion resistance of some ferritic stainless steels used to make knives compared to the good corrosion resistance of austenitic stainless steels used for forks and spoons. The structural change in the stainless steel is caused by increasing the amount of nickel in the steel by only a small percentage.

If the corrosion is controlled by the anodic reaction in a galvanic couple, minor changes to the cathode material may not be as important as differences in the anode

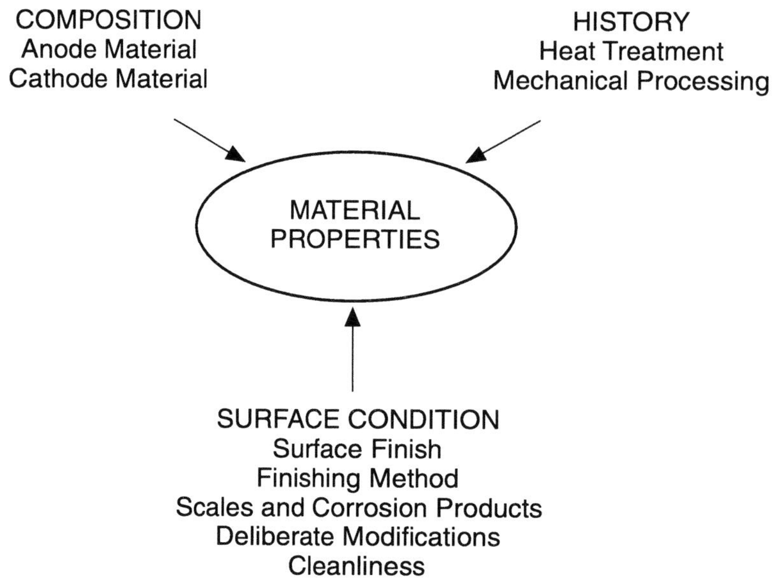

FIGURE 2.2 - Material properties affecting galvanic corrosion.

material. For example, a small piece of zinc will corrode at the same rate whether it is coupled to a very large piece of steel or of bronze in seawater. Here, the corrosion rate is limited by how fast the zinc can corrode. The reverse may be true for corrosion controlled by the cathodic reaction. For example, the galvanic corrosion rate of a large piece of zinc coupled to a small piece of bronze in seawater may be similar to the galvanic corrosion rate of a similarly sized piece of steel coupled to the bronze. In this case, the increase in corrosion is due to the cathodic reaction on the bronze, independent of the type of material corroding.

Thermal history can also have a large effect on corrosion performance. Heat treatments can affect the alloy structure and can result in certain elements becoming nonuniformly distributed in the alloy, which can change corrosion performance. For example, the distribution of tin in certain sacrificial aluminum anode materials can be changed by heat treating the anodes, changing the anode efficiency by more than 20 percent. Mechanical processing such as rolling of steels can also change corrosion test results. For example, heavily rolled stainless steels are cathodic to the same steels with no rolling. Materials being tested should have undergone the same thermal and mechanical processing as the material used in service.

Surface condition includes a variety of important factors. Examples include surface finish (roughness), and whether the surface was prepared by machining or grinding. Some surface preparations, such as machining or shot blasting, harden the surface; others, such as chemical etching, may cause certain elements to be removed from the surface faster than others. It may be best to leave the factory-produced scale on the surface or otherwise use a specimen with a surface left as produced if that is the way the material will be used in service. Heat treatments may affect the surface, as when carbon is lost from the surface layers during high-temperature treatments of steel in air, or when lubricants used in forming the material are oxidized to carbon during later heat treatments.

Sometimes, deliberate changes are made to the surface to change its corrosion performance, for example, acid treatments to remove surface contamination and lower the corrosion rate of stainless steels, or anodic oxidation of aluminum to form a protective surface layer. Finally, residual oil or grease from forming or machining operations or from fingerprints may be important in some laboratory tests. Oil from tube-rolling operations has been found to affect significantly the corrosion performance of copper-nickel alloy tubes in seawater, for instance. If there is doubt about whether residual grease will have an effect, chemical cleaning or degreasing of

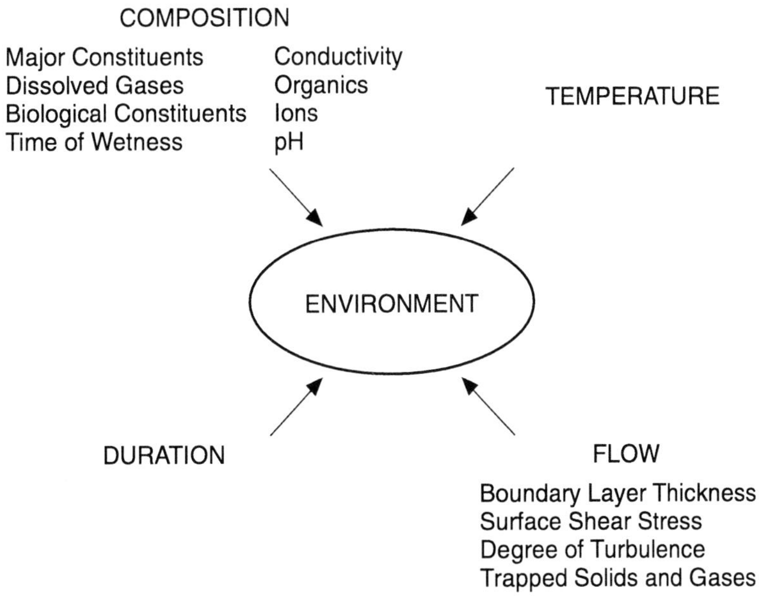

FIGURE 2.3 - Environmental factors affecting galvanic corrosion.

the surface may be necessary before testing. ASTM Standard G1 is a set of standard procedures for chemical cleaning to prepare a surface for corrosion testing.[1]

Environment

It is important to reproduce the service environment closely. Small changes in environment that affect conductivity or change the concentration of inhibitors or depassivators for the reactions can cause large differences in galvanic corrosion. Environmental factors such as temperature, pH, or conductivity may be changed to speed up a galvanic corrosion test, but this must be done with care because increasing the rate of the reactions may change the reaction mechanism, giving results that should not be used in the service situation. The composition of the environment should normally be made as close to that of the service environment as possible. This includes major constituents such as salt in seawater, dissolved gases such as oxygen, minor constituents such as pollutants, ions that affect conductivity, dissolved organics like amines, and even biological organisms such as sulfate-reducing bacteria. Temperature, flow, and some atmospheric factors can also affect galvanic corrosion. These are illustrated in Figure 2.3 and described below.

Composition

Some ions in solution can have a major effect on galvanic corrosion. The chloride and other halogen ions like bromide are powerful pitting agents for stainless steel and nickel alloys. Sulfate ions may also have this effect. Most charged ions will have some effects on metal surfaces; phosphates can be corrosion inhibitors for steels. Many substances will affect pH, which is important when the cathodic reaction is the reduction of hydrogen ions to form elemental hydrogen. Thus, a lower pH may stimulate this cathodic reaction, which, in turn, may create another problem: the hydrogen generated can diffuse into the cathodic metal to cause cracking.

Dissolved Gases

The most important dissolved gas in a corrosion test is usually oxygen, because it is frequently part of the major cathodic reaction. Some pollutants, such as hydrogen sulfide, may also be part of the cathodic reaction or may act as a poison or catalyst to change the rate of the anodic or cathodic reactions. Carbon dioxide or sulfur dioxide may also affect pH. Dissolved hydrogen can cause cracking in metals only under many atmospheres of pressure.

Minor Constituents

Very small amounts of some substances can affect galvanic corrosion. For example, the sulfide ion in concentrations of only five to 10 parts per billion can greatly increase the rate of galvanic corrosion in seawater and can even reverse the direction of galvanic current. An instance of this is the reversal in galvanic current between 70-30 copper-nickel and tin-bronze in seawater caused by a few parts per billion of hydrogen sulfide. Equally small amounts of heavy metal ions such as copper or mercury can greatly increase the corrosion of aluminum. For this reason, copper antifouling paints are not used on aluminum-hulled boats.

Sometimes the local environment can contain different minor constituents than the larger surroundings. This can occur when corrosion products from nearby structures fall onto or run off toward the structure of interest. For example, rust from an iron structure falling onto a nearby aluminum structure can cause local galvanic corrosion effects. Bird droppings have been known to create strange corrosion effects. Fertilizer or other runoff from the shore of a river may make the composition of the water near the riverbank quite different from that of the water midstream.

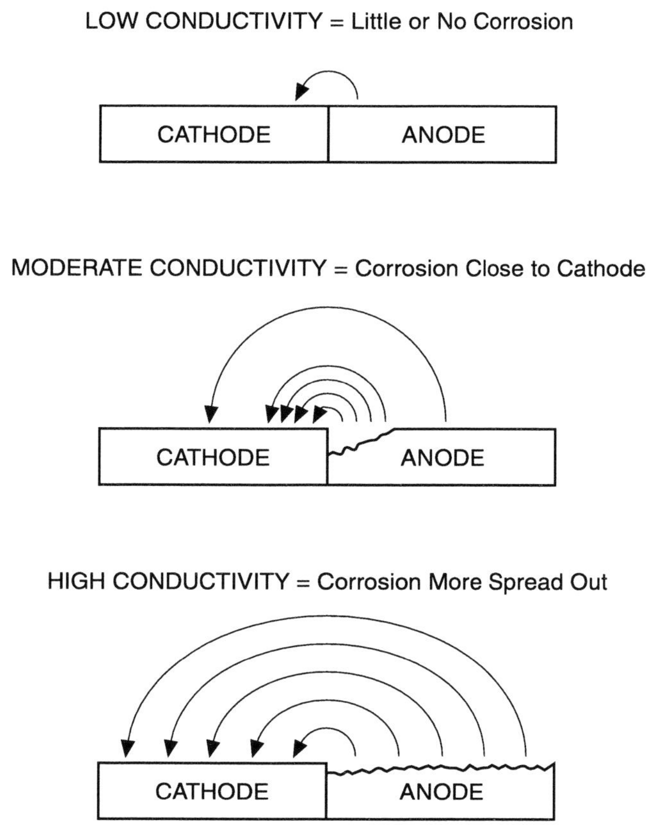

FIGURE 2.4 - Effect of electrolyte conductivity on magnitude and distribution of galvanic corrosion.

Conductivity

Conductivity of the solution influences the amount and distribution of galvanic corrosion, as shown in Figure 2.4. In pure, de-ionized water, galvanic corrosion is usually slight because of low conductivity, which results from the high resistance to ionic current flow through the water. Even small amounts of charged ions can increase conductivity just enough to cause a corrosion problem. Thus, adding a small amount of salt to tap water sitting in a stainless steel sink may cause galvanic corrosion near the drain, where different materials are used. In moderately conductive solutions, galvanic corrosion will occur but will have its greatest effect near the junction between the different materials making up the galvanic couple. In highly conductive solutions such as concentrated acids or seawater, the total galvanic corrosion will be greater and will be more evenly distributed over the anode surface than in poorly conductive solutions.

Organics

Organic substances can affect galvanic corrosion in the same ways as inorganic substances: participating in reactions, inhibiting reactions, and changing conductivity. Also, many organics react strongly with metal ions to increase corrosion—for example, certain amino acids corrode copper. Others are powerful inhibitors—for example, benzotriazole protects copper. Some organics may even physically shield the surface. This can occur, for instance, in water polluted by oil if the oil coats the surfaces of the materials in the galvanic couple.

Bio-Constituents

Some environments, such as seawater, support a great deal of biological activity. In these environments the activity of both visible and microscopic fouling organisms can affect galvanic corrosion in a number of ways. Barnacles can shield surfaces by their attachment and growth. Bacteria on a surface can consume substances, such as oxygen, that affect corrosion reactions, or generate substances, such as organic acids, that affect corrosion. For example, in aerated seawater under slime deposits, sulfate-reducing bacteria near the surface can generate potentially harmful sulfides that would not normally be found in the surrounding seawater. These same bacteria have been known to cause corrosion problems in the downhole environment in oil drilling. Bacteria are known to affect the corrosion potential of metals in seawater. Positive potential shifts of up to 500 millivolts on noble metals like titanium or nickel in seawater have been attributed to the presence of bacteria. Such large potential shifts could cause reversals in galvanic corrosion behavior, making an anode into a cathode when the bacteria are active.

Temperature

Temperature controls the rate of anodic and cathodic reactions, the rate of movement of atoms and ions, solubility of dissolved gases and corrosion products, and conductivity of the electrolyte. Temperature should be maintained as close as possible to that in service. If the temperature range is moderate (0 to 100° C), this often is done by using a constant-temperature bath into which the test cell is placed. For higher temperatures, or pressures above atmospheric, special corrosion cells called autoclaves frequently are used. When using such cells, the test specimens or assembly should not come into electrical contact with the cell, because these cells are usually metallic and would have an unwanted galvanic effect if allowed to contact the test assembly.

Flow

Flow of the fluid can be one of the most important influences on galvanic corrosion, and yet it is by far the most difficult parameter to duplicate. Usually, just duplicating the flow speed (for example, in meters per second) is not enough for good flow modeling. Recent evidence strongly suggests that, at least for some materials in saltwater, the degree of turbulence controls corrosion rate. For other materials or environments, corrosion could be controlled by diffusion of molecules or ions through a stationary fluid layer at the surface, in which case the thickness of this layer should be matched with that in the service environment. The cathodic reaction prominent on most metals in neutral or alkaline waters is oxygen reduction, which is controlled by the rate of

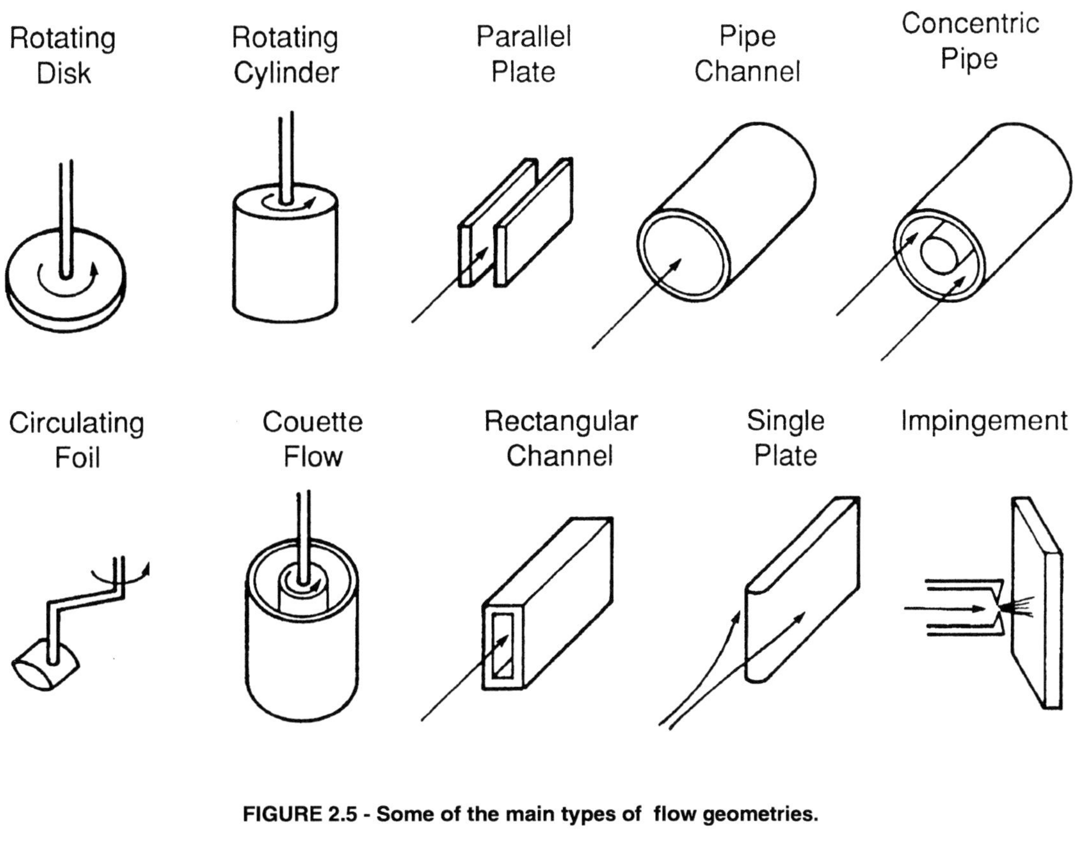

FIGURE 2.5 - Some of the main types of flow geometries.

diffusion of oxygen through a fluid boundary layer. For cathodic reactions in water, therefore, the tests should approximate the thickness of this layer, called the Nernst boundary layer.

Corrosion could be controlled by diffusion through a corrosion product layer on the metal surface. This corrosion product layer will form as a function of the degree of turbulence or shear stress at the metal surface generated by the fluid flow. Again, these parameters should be matched to those expected in service. Calcareous deposits, which are formed on a cathodically protected metal in seawater, act as a barrier to the diffusion of oxygen and therefore control the cathodic reaction. The type of flow, turbulent or laminar (smooth), must be duplicated, particularly if ion movement is important in the corrosion process. Trapped gas bubbles or solids can cause scouring effects that may need to be considered. Unfortunately, in most cases the mechanism for the corrosion reactions is not known, making flow modeling difficult.

The number of different types of flow geometries is probably as great as the number of experimenters, and some of the main types are shown in Figure 2.5 (see Page 15). Each type has advantages and disadvantages. A rectangular flow channel allows for easy mounting of the specimens parallel to each other, but certain hydrodynamic parameters such as boundary layer thickness are difficult to calculate and vary with position on the specimen faces.

Flow is well characterized in the rotating disk for laminar flow and rotating cylinder for turbulent flow devices, but specimen mounting is difficult and the different materials must be adjacent, not facing, leading to galvanic corrosion that is different at specimen edges. Channels with circular cross-sections may be made of the test materials but are difficult to construct unless the test material is piping or tubing. The circulating foil apparatus is expensive to make, and each of the other apparatus types illustrated has unique advantages and disadvantages.

One common concern is the mounting of specimens to prevent unwanted crevices from forming and to keep electrical connections dry. O-ring seals prevent leaks well but form bad crevices. Specimens may be mounted in epoxy or another nonmetallic mounting material, but the bond to the metal must be excellent to prevent the formation of a crevice and to prevent leaking, especially over long-duration tests. Sometimes unexpected problems occur with flow cell materials, such as stress corrosion of some clear plastics in seawater.

Atmospheric Variables

Environmental composition can take on a whole new meaning when applied to atmospheric testing. Minor constituents, such as SO_2, NO_2, and chlorides from sea atmospheres, are sometimes known as pollutants. Some environmental variables have meaning only for atmospheric tests. The most important of these is time of wetness, which is determined by the frequency and duration of rainfall, dew point, corrosion products and solid deposits on the test specimens, and specimen temperature. Thick specimens may actually experience a longer time of wetness during seasons when dew is prevalent than thin specimens because thick specimens may take longer to heat up in the morning sun to evaporate the condensed moisture. ASTM Standard G84 gives help on building and using time-of-wetness sensors.[2]

Test specimen placement and geometry are also important in atmospheric galvanic corrosion tests. Test results may be affected by the height of the specimen above ground; angle of exposure from the vertical; and by the direction relative to prevailing winds, nearby sources of pollutants such as factories or the sea, or the sun. The extent of ultraviolet (UV) radiation from the sun can be changed by specimen shape, angle to the ground, or shading. The alternating wet-dry cycles of atmospheric corrosion can cause concentration of minor atmospheric constituents. Water drainage and trapping in crevices should also be considered, as shown in Figure 2.6. Much good information about the important characteristics of atmospheric corrosion sites may be found in ASTM Standard G91 for "Characterization of Atmospheric Test Sites."[3]

Geometry

The size of the test specimens can affect the relative amounts of edge and surface, so that small specimens will emphasize corrosion phenomena that are different at specimen edges while large specimens will de-emphasize these effects. Large specimens must be placed in large test cells. This will prevent corrosion products from building up to unacceptable levels in the test solution. The size of a specimen will also affect the occur-

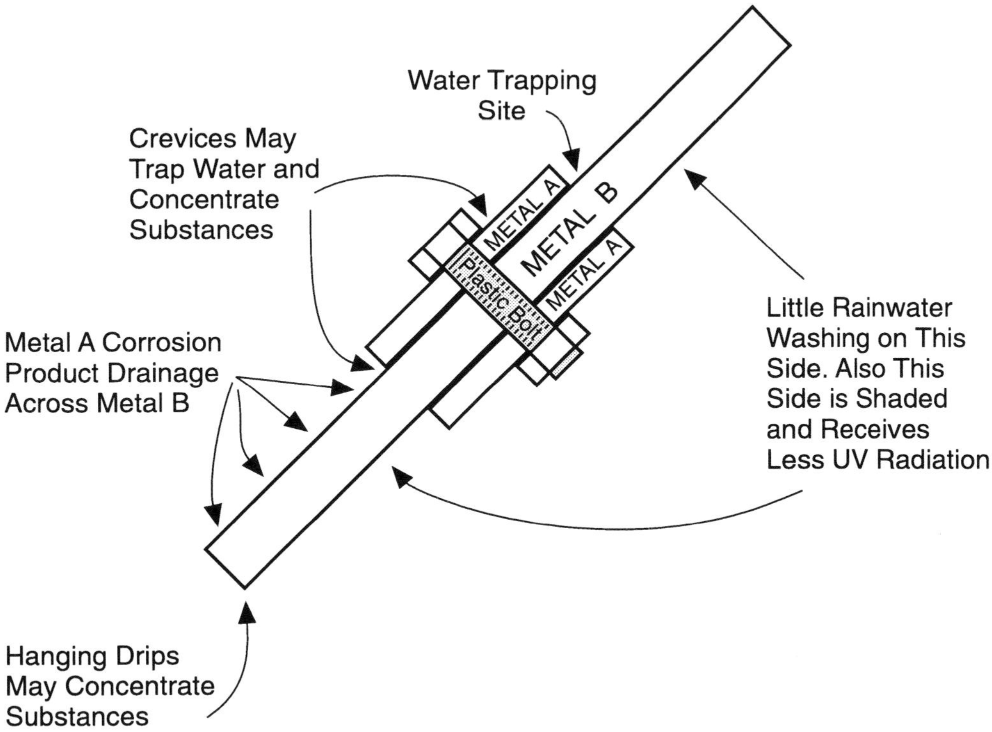

FIGURE 2.6 - Water damage on a galvanic corrosion test specimen (cross section).

rence of some localized phenomena such as pitting. A large specimen will have a greater statistical chance of having a pit initiate—and a greater chance of having a deep pit—than a small specimen. If corrosion has a small probability of occurrence, then a larger specimen is more likely to experience it. Such effects may be described and analyzed by the statistics of extreme values, which can be found in advanced texts on statistics.

Specifics for the Forms of Corrosion

Each form of corrosion that may be increased by galvanic effects has its own variables, which must be simulated in the test in order to duplicate adequately the conditions in service. For stress corrosion, the magnitude (amount) and state (direction) of stress must be simulated. Crevice corrosion depends on the geometry of the crevice, with tighter crevices usually leading to quicker initiation of corrosion. More details about the specific forms of corrosion are given in the other volumes in this series.

Duration of Test

The duration of a galvanic corrosion test should be long enough to allow conditions to develop the way they do in service. Galvanic corrosion is frequently time dependent in unpredictable ways. If the desired lifetime of a component is many years, it often is impractical to test for that length of time. In this case, some method should be used for following the corrosion process until steady state—from which extrapolations to long-term service may be made—is reached. If seasonal variations are expected, as in atmospheric corrosion exposures, then a good exposure period is some exact multiple of a year so that each season is experienced the same number of times. Sometimes the best that can be done is to run the test until a decision must be made or until the testing budget is exhausted. Fortunately, in some cases where corrosion products are soluble, the steady state may be reached quickly.

Replication

Replication is the use of several identical test specimens exposed under identical conditions. A well-planned test should consider the need for replication. Statistical methods for planning of exposures are available to minimize the number of specimens. Because corrosion tests often are variable in nature, it is usually advisable to perform each test at least in duplicate—and sometimes in triplicate or more—to get a feel for the variability in behavior. The results from replicated tests can be valuable when analyzed by statistical procedures. Further discussion of this is beyond the scope of this volume.

References

1. ASTM G1-90, "Standard Practice for Preparing, Cleaning, and Evaluating Corrosion Test Specimens." (Philadelphia, PA: ASTM, 1990).

2. ASTM G84-89, "Standard Practice for Measurement of Time-of-Wetness on Surfaces Exposed to Wetting Conditions as in Atmospheric Corrosion Testing." (Philadelphia, PA: ASTM 1989).

3. ASTM G91-86, "Monitoring Atmospheric SO_2 Using the Sulfation Plate Technique, Practice for." (Philadelphia, PA: ASTM 1986).

Bibliography

General

Bimetallic Corrosion. Publication No. 14. National Corrosion Service, Department of Industry, England: Colibri Press Ltd., 1982.

Champion, F.A. *Corrosion Testing Procedures.* London, UK: Chapman and Hall, 1952.

Fontana, M.G. *Corrosion Engineering.* New York, NY: McGraw-Hill, 1986.

Haynes, G.S., R. Baboian. "Reducing Galvanic Corrosion with Transition Metals." *Materials Performance* 16, 6 (1977): pp. 36-39.

Rowe, L.C., R.L. Chance. "Fundamentals of Corrosion Testing." *Automotive Corrosion by Deicing Salts.* Ed. R. Baboian. Houston, TX: NACE, 1981.

Shreir, L.L., ed. "Metal/Environment Reactions." *Corrosion*, Vol. 1. London, UK: Newnes-Butterworth, 1976.

Sprowls, D.O. "Planning and Preparation of Corrosion Tests." *ASM Metals Handbook*, Vol. 13, 9th ed. Materials Park, Ohio: ASM International, 1987.

Tomashov, N.D. *Theory of Corrosion and Protection of Metals.* New York, NY: Macmillan Company, 1966.

Uhlig, H.H. *Corrosion and Corrosion Control.* New York, NY: John Wiley and Sons, Inc. 1971.

CHAPTER 3

Laboratory Testing

Standardized Tests

Because laboratory testing for galvanic corrosion involves so many variables, there are no standardized laboratory testing procedures. If another form of corrosion is being studied, then a slight modification of a standardized test for that corrosion form may be used. For example, to study galvanically accelerated stress corrosion, one may use ASTM Standard G30 on stress corrosion testing and add a wire between the specimen and a second material to create the galvanic couple.[1] To study the breaking strength of a piece of metal, a standard mechanical pull test such as that found in ASTM Standard Test Method E8 may be used.[2] To study the effect of galvanic corrosion on breaking strength, the same test may be used on pieces of metal after exposure to the environment of interest, both with and without galvanic coupling to a second material. Subtracting the strength of the metal with galvanic coupling from that without it will give the effect due to galvanic corrosion.

There are two guidelines for galvanic corrosion testing published by ASTM. ASTM Standard Guide G71 on "Conducting and Evaluating Galvanic Corrosion Tests in Electrolytes" is the basis for much of this book.[3] Procedures detailed in ASTM Standard G82 on "Development and Use of a Galvanic Series for Predicting Galvanic Corrosion Performance" will be described in more detail later in this chapter.[4]

Accelerated Tests

The key to accelerated galvanic corrosion testing techniques is simple: Avoid them if possible, otherwise use them with care. Usually, any procedure that will accelerate the rate of galvanic corrosion will also change the mechanism of that corrosion, making the results invalid when applied to a service situation. For example, galvanic corrosion of steel alloys coupled to bronze may result in steel being oxidized to the Fe^{2+} ion. Acceleration of this effect by the use of an external current to force higher corrosion rates may result in the formation of the Fe^{3+} ion instead. This could affect the protective nature of the corrosion products, changing the very nature of the corrosion process being studied.

Methods of accelerating galvanic corrosion tests include changing area ratio, varying composition of the solution, and imposing currents from external sources. Unless it is certain that the mechanism remains unchanged, accelerated testing is not advisable. Galvanic corrosion information can be obtained fairly quickly by the electrochemical test techniques described below. These tests should be designed, not to accelerate the rate of corrosion, but to measure accurately the small amount of corrosion as it occurs. This will enable the short-term corrosion rate to be used to make judgments about long-term service, provided that the corrosion rate has stabilized during the time frame of the test. If accelerated testing must be performed, particular care should be given to interpreting the data in light of the service situation.

Common Issues

There are several things that are common to all laboratory testing for galvanic corrosion. All specimens must be mounted and held in position in the test cell, and an electrical connection must be made to allow current to flow between the materials of interest. Some test cell considerations are common to all test techniques. These include methods of establishing flow, and cell size. These subjects will now be discussed in detail.

Mounting and Electrical Connection

Sometimes the metals in a galvanic cell are mounted together, with electrical connection made by physical contact. Other times they are mounted separately and connected with a wire that frequently is isolated from the test environment. In the discussions below, methods of mounting the specimens separately are covered first. At least two such mounted specimens will be required in any galvanic corrosion test.

Any mounting procedure may produce a crevice. This is an area where the device that holds the specimen touches the specimen and limits access of the test solution to that part of the surface. Because the movement of reactants and reaction products to and from such crevices often is not typical of what happens in the areas away from the crevice, corrosion in crevice areas may be quite different from what occurs in the service situation being simulated. For instance, if stainless steels are involved in the galvanic cell, there is a chance that crevice corrosion will occur during the galvanic corrosion test; this will complicate the galvanic corrosion analysis. Crevice-free assemblies can be made when an extremely tight seal is made by some gasket devices (although this is open to debate) and when a strong chemical bond is made between the specimen and a nonmetallic mounting material such as epoxy. In both of these cases, extreme care must be taken to ensure sufficient tightness of the seal or bond to prevent water

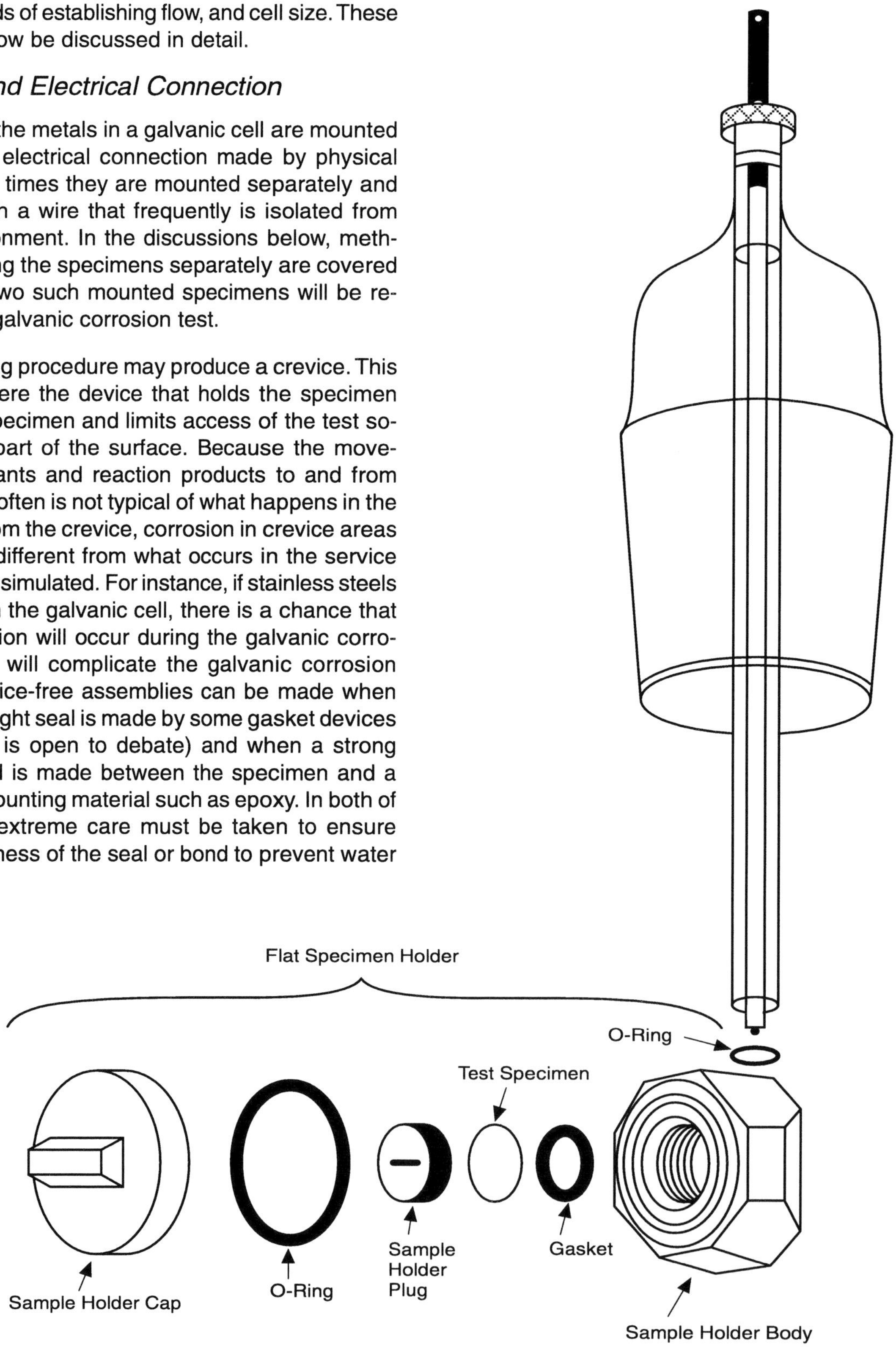

FIGURE 3.1 - Gasketed specimen mount (Courtesy of EG&G Instruments - Princeton Applied Research).[5]

from getting between the specimen and the mount. Neither of these methods will unequivocally eliminate crevices. While great effort is required to minimize crevices in a laboratory setting, in the field it is almost impossible to avoid them completely.

To simulate service corrosion, crevices may be desirable in a laboratory test. In that case, the type of crevice—size, material forming the crevice, tightness, and so on—should be similar to that in service. Mounting must always be accomplished in such a way as to allow electrical connection to the specimen. Either the electrical connection must involve the same materials as the specimens to avoid adding materials to the galvanic couple, or the connections must be kept dry. Several examples of common specimen mounting procedures are described below.

Gasketed Mounting

Figure 3.1 shows a gasketed type of specimen mount that is commercially available (there are only minor differences among mounts from several manufacturers). The specimen is a flat disk that is pushed against a circular gasket or washer by pressure from the back, usually supplied by a threaded plug or cap. Only the specimen face inside the circle of the washer is exposed to the test solution; the washer keeps the edges and the back of the specimen dry. The gasket is usually made of a material that is not easily wetted by the test solution to minimize the possibility of the area between it and the specimen becoming wet. Typical materials include fluorocarbons such as TFE fluorocarbon and rubbers such as Buna-N†. O-rings are usually not used as they can create a severe crevice situation.

The gasket holder is usually made of a nonmetallic, nonconducting material such as nylon, TFE fluorocarbon, polyvinylchloride, glass-reinforced plastic, epoxy, Roulon†, or Micarta†. Particular care should be taken in selecting the materials for highly stressed areas such as threads. Some of these materials may deform slowly and continuously under stress, especially at temperatures well above room temperature. For example, some types of TFE-fluorocarbon are known to deform enough at room temperature to cause a mount to loosen during the course of the test. It may be desirable to have a gasket material that deforms slightly, because this will seal any irregularities such as scratches in the specimen surface. Specifications for the seal materials involved must be consulted to determine the temperature above which appreciable deformation may occur.

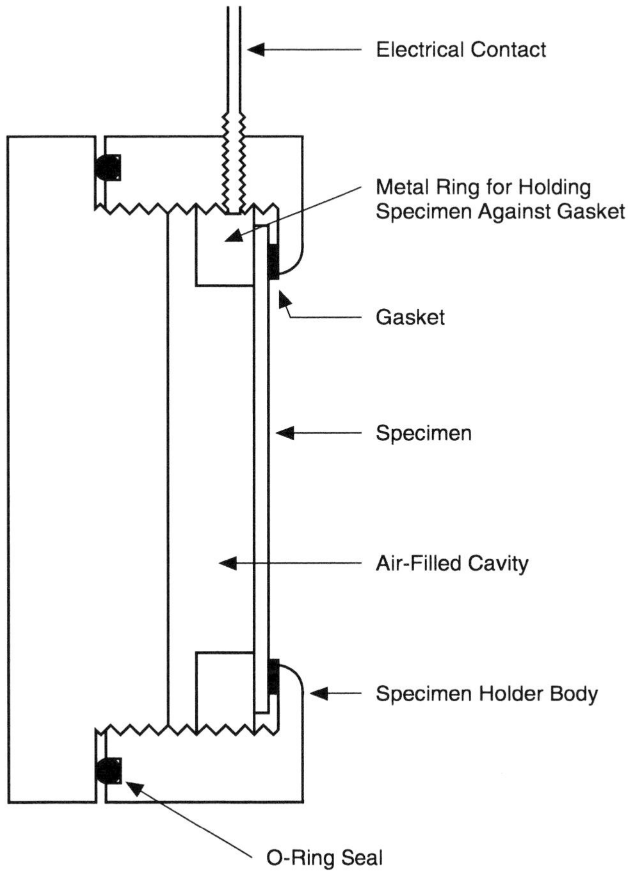

FIGURE 3.2 - Cross section of specimen holder for thin sheets.

The electrical connection in gasketed holders is usually made on the back face of the specimen (see Figure 3.1), although at least one company makes its electrical contact on the edge. Electrical connection can be made by a threaded screw contacting the center of the back specimen face. The screw must be tightened sufficiently to make a good connection but not enough to bend the specimen, as this may affect the watertight seal. This can be a problem with thin specimens. Another way to make the electrical connection is by a threaded ring on the back side, as shown in Figure 3.2. This is a better procedure for thin specimens because it provides less contact force per unit area. The back of specimens using this type of holder should be thoroughly cleaned before assembly to remove nonconductive deposits.

† Trade names

If the specimen is not held tightly enough in a gasketed holder, leakage will occur that will cause the electrical connection point to get wet. This is undesirable for two reasons. First, the contact screw is an additional metal that is introduced into the galvanic cell. Second, corrosion at this point may eventually result in a loss of electrical connection. If leakage is discovered at the conclusion of the test, it usually necessitates a rerun.

The gasketed type of specimen holder requires that the exposed specimen face be prepared so it presents a smooth face to the gasket. Therefore, it cannot be used for specimens with rough surfaces such as intact scale from heat treatments at the producer's mill or other deposits. Typical preparation involves grinding and polishing at least through 600-grit, and better seals may be obtained by polishing down to a mirror finish.

Air entrapment is an important problem with this type of holder, as shown in Figure 3.3. The specimen should never face downward, because air will tend to be trapped as the specimen is immersed and any gases evolved in the reactions may also be trapped. If the specimen faces upward, on the other hand, corrosion products will settle on the surface. For these reasons, this holder is usually used with the specimen face vertical. Even in this configuration, care should be taken during insertion of the holder into the solution to avoid trapping air in the upper part of the gasket area.

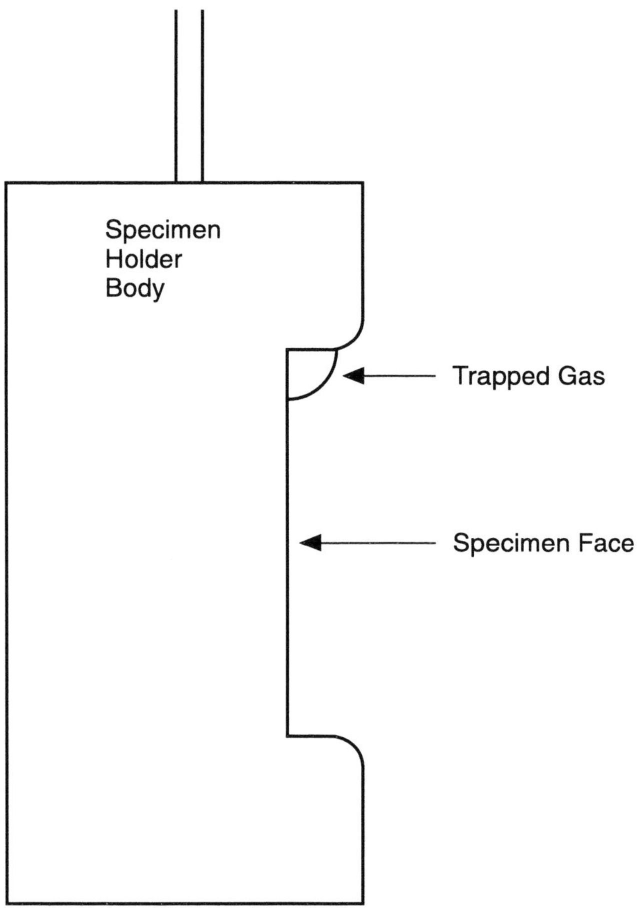

FIGURE 3.3 - Trapped gas in specimen holder.

Rod Mounting

A second popular type of specimen holder is shown in Figure 3.4. This holder requires a chunk of material for the sample. This chunk is usually machined into a cylinder, although this is not essential. A small area on the upper surface of the specimen is ground flat and a threaded hole is tapped partway into it. A threaded rod is screwed into this hole and serves as both a mounting device and an electrical connection. The rod is kept dry by a nonmetallic tube, usually glass, with a special conical gasket at the bottom. A nut at the top of the threaded rod applies enough downward pressure on the tube to cause the gasket to seal. The potential crevice area in this type of holder is small, and the gasket pressure is high. This makes this type of holder a good choice for minimizing the possibility of crevice corrosion. Specimen surfaces need not be prepared except where in contact with the gasket. Cylindrical specimens can usually be surface ground if desired by putting a short threaded rod in a drill and threading the specimen onto the rod. The drill will spin the specimen nicely for sanding the perimeter and bottom end, and the top end may be sanded after removal from the drill.

Because of high pressure on the gasket, deformation usually occurs with this holder, and gaskets should not be reused. Overtightening can cause leaks, as the gasket may deform so much that it will no longer seal properly. This is shown in Figure 3.5.

Partial Immersion Mounting

A third popular mounting procedure, shown in Figure 3.6, allows a portion of the specimen to extend above the test solution. The portion of the specimen at the water line is usually made small to minimize waterline effects on the test results. Above the solution, clamps may be used to hold the specimen, and the electrical connection made to this point need not be specially treated. This works well for quick tests, but during longer tests, vapors from the test solution can corrode the

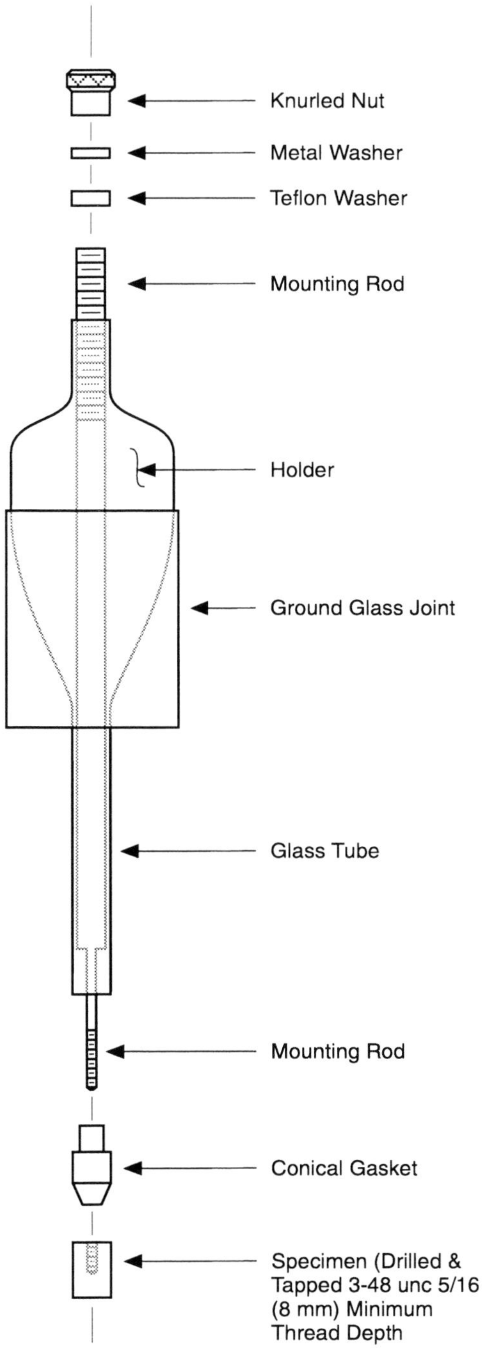

FIGURE 3.4 - Rod mount.

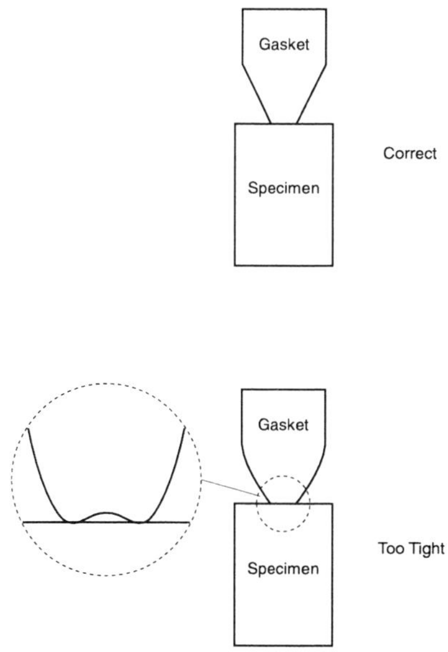

Figure 3.5 - Conical gasket compression.

Wall Mounting

Specimens may be mounted in the wall of the test cell with appropriate gasketing, as shown in Figure 3.7 (see page 27). This is particularly useful for flowing exposures because this arrangement can be designed without a step in the wall profile (a step could cause flow disturbance). Electrical connection can be made to the electrical contact or the clamps. This is the only mounting procedure that introduces no crevice area. However, a waterline area is present instead, and it can be an initiation site for corrosion. If waterline effects are expected to be significant, then another mounting procedure should be used.

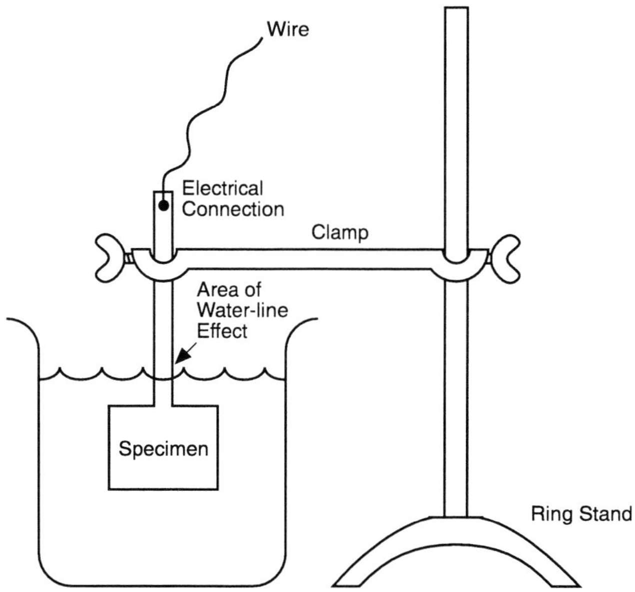

FIGURE 3.6 - Partial immersion mount.

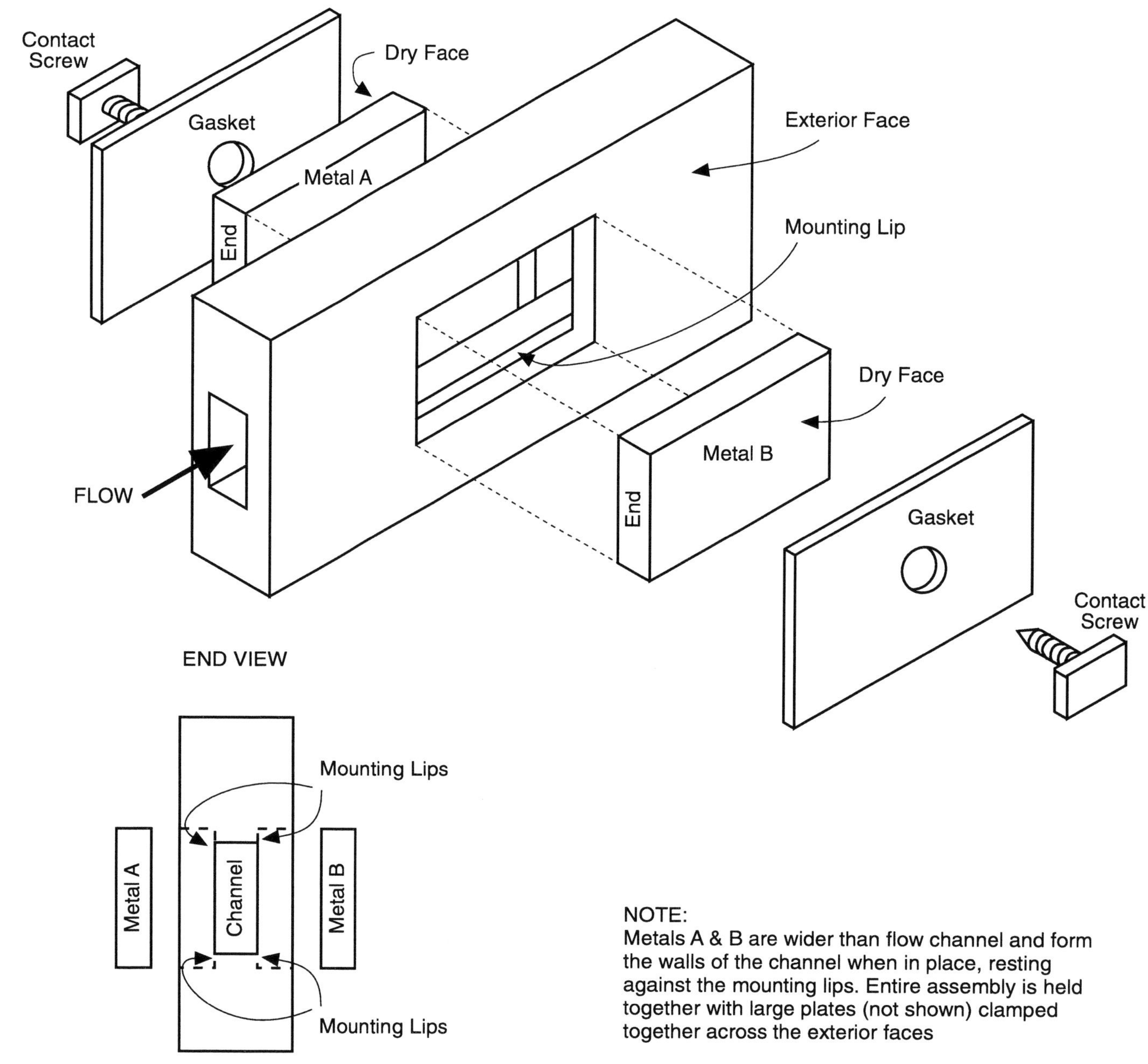

FIGURE 3.7 - Wall mount.

dry specimen back outside the cell. This type of mounting is prone to leakage and to crevice corrosion because of the complexity of the seal. Leakage may be minimized in the configuration in Figure 3.7 by using gasketing or sealant on the upper and lower edges of the specimens where they contact the indentations that form the top and bottom walls of the flow channel. The ends of the specimen are much harder to seal, and this is why the gasket behind the specimen is required.

Epoxy Mounting

Figure 3.8 shows that it is possible to make an electrical connection to the back of a specimen and then mount it in epoxy or a similar mounting material. The wire con-

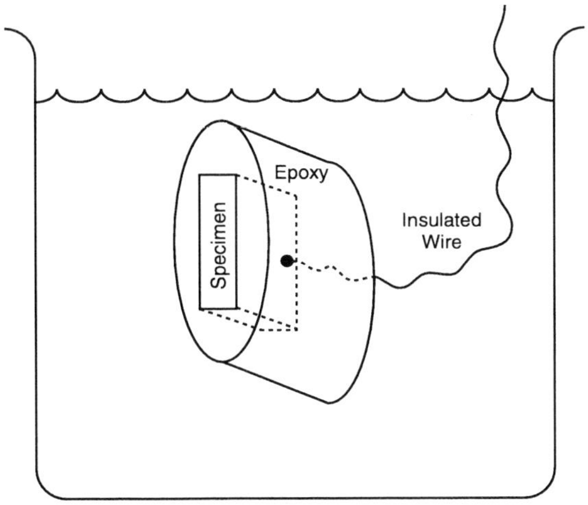

FIGURE 3.8 - Epoxy mount.

nected to the specimen should have insulation that is resistant to the test solution. This insulation should terminate inside the epoxy mount to maintain the seal. The resultant assembly can then be mounted in a variety of ways: Used as the wall of a cell, mounted in the wall, or suspended in the test solution. This type of mounting may create a crevice around the specimen, the severity of which is highly dependent on the type of mounting material used. For seawater, some epoxies have been found to work extremely poorly, while others work so well that they cannot be removed for post-test mass-loss measurements. Specimen surfaces must be reground after mounting if a smooth transition between specimen and mount is desired, for example, for flow tests. However, such regrinding will invalidate any mass measurements made before mounting.

Press-Fit Mounting

Press-fit holders, as in Figure 3.9, are perfect for complex geometries that require a flush fit, which is advantageous for certain types of flowing tests such as one using the rotating disk apparatus illustrated. This type of mounting must be manufactured to exacting dimensions using materials that will not continue to deform after specimen insertion. Specimen holders may not be reused very many times before the tight seal is lost, and they will not work for materials that are susceptible to crevice corrosion. It is difficult to prevent water leakage to the electrical connection on the back of the specimen unless the specimen is mounted using a sealant, which makes specimen removal more difficult. For the configuration in Figure 3.9, specimens may be removed by tapping them from the back side with a rod inserted through the threaded hole in the holder.

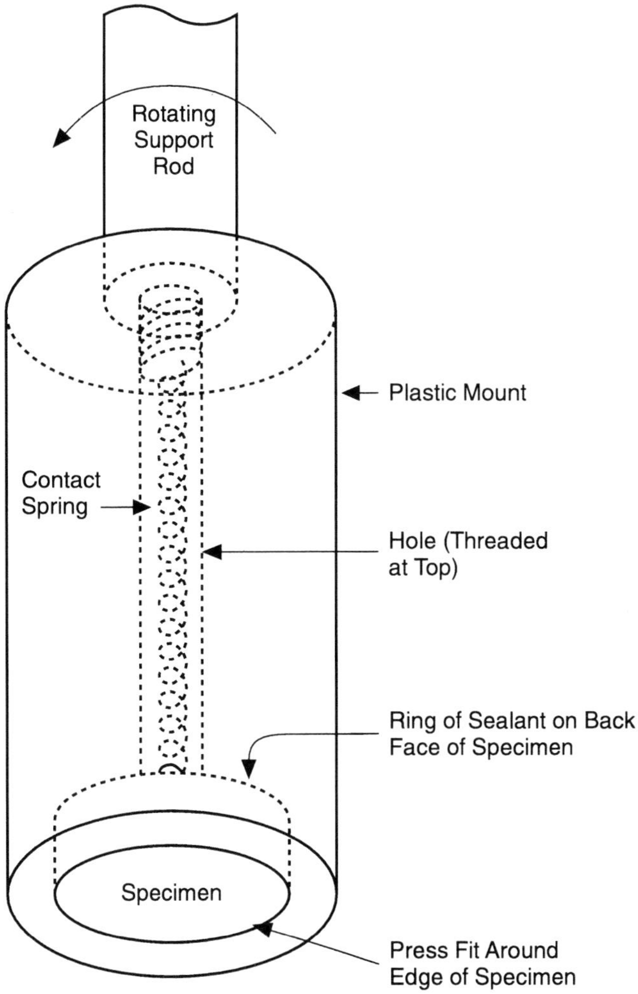

FIGURE 3.9 - Press-fit mounting.

Full Immersion Mounting

Perhaps the simplest mounting procedures, shown in Figures 3.8 and 3.10, are to place the specimen in a nonmetallic holder, to set it on the bottom of the test cell, or to hang it by the wire connection. The electrical connection must be secure, the wire must be thick and strong with insulation resistant to the test solution, and the connection area must be sealed to prevent wetting. In seawater, silicone sealants have been found to be adequate for this purpose for short exposures, while epoxies or coal tars are preferred for longer exposures. Commercial stop-off lacquers and other materials that will sometimes work adequately for short tests are sold by electroplate suppliers. The disadvantages of this mounting method are the lack of solidity of the mount and the need for a sealant, which may cause crevice corrosion.

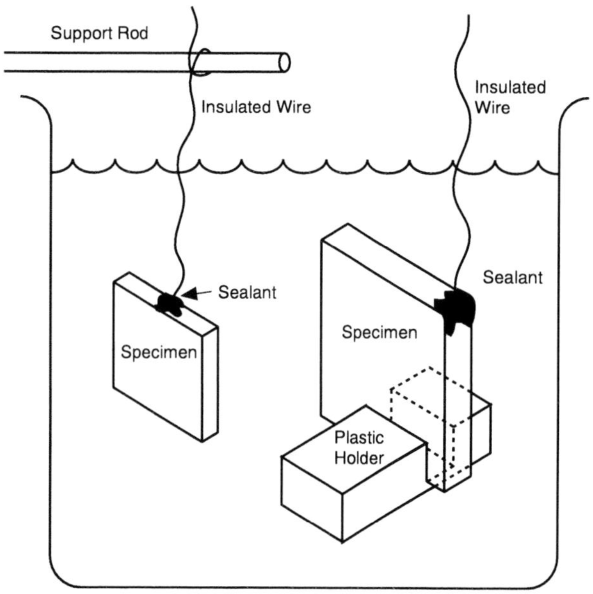

FIGURE 3.10 - Full immersion mounting.

Other Mounting Methods

There are perhaps as many mounting procedures as experimenters. A few of the more novel ideas are shown in Figure 3.11. These include making tubular specimens into the walls of the test cell for flow tests, using etched printed circuit boards made of the specimen material, and using the same loading bolts that hold the crevice formers in a galvanically accelerated crevice corrosion test.

Connecting the Wire

Wire insulated with a nonconductive coating such as plastic is required wherever the wire is exposed to the test solution. Insulation may not be needed in certain cases where it can be determined that current to or from the wire will not affect the test results. This would occur if the area of the exposed wire is so small that its reaction current does not significantly add to the total current flowing through it. If insulation is not used, the wire should be made of a material that will not deteriorate during the test, such as platinum or tantalum in seawater. The wire material should also have a corrosion potential in the test solution that is similar to, or more positive than, the most positive specimen in the test. For example, a couple in seawater between steel (-650 mV) and bronze (-200 mV) could use an uninsulated stainless steel (+100 mV) wire but not an aluminum (-1100 mV) wire. The aluminum wire has the most negative potential of the three materials and could fail prematurely by galvanic corrosion.

Connecting a wire to a piece of metal is more difficult than it may seem initially. A rule of thumb is always to make the connection better and more secure than would seem necessary at first sight, and this is even more important for tests that will take more than 30 days. Some methods are shown in Figure 3.12 (see Page 30). Wires may be soldered to copper or to iron-based metals, or brazed to other materials, provided the heat does not introduce unwanted changes in the material. To determine this, the detailed metallurgy of the alloy must be known. Sometimes spot-welding is better because it is fast and introduces little heat damage, but the connections are usually weak. Ductile materials such as silver may sometimes be connected to similar materials by cold-welding (placing the wire on the specimen and hitting it with a hammer). A drilled hole may have a wire inserted and then the hole edges peened (deformed with a sharp tool and hammer) in around the wire. This method is weak and provides an unreliable electrical connection, but it can be used for materials that cannot be soldered or brazed to and that will not hold a bolt well because they are too weak, for example, low-strength aluminum alloys. A screw or a threaded rod in a threaded hole can make a good connection, particularly if the wire is secured under the screw with a wire lug rather than by wrapping it around the screw before tightening. If the specimen cannot be threaded, a hole can sometimes be drilled through it, and a nut and bolt used for connection. When the screw material and specimen have different rates of expansion with temperature, electrical connections can loosen over time as the temperature changes cause relative movement. Combinations of different materials can generate extremely small voltages through thermocouple effects caused by differential heating of dissimilar metals. If voltages must be measured to an accuracy greater than 1 mV in the test, thermocouple effects between the attachment wires and the specimens must be considered, but for most electrochemical measurements the error introduced is insignificant.

The method of wire attachment may affect the accuracy of mass-loss measurements. For example, solder or brazing material may not be able to be removed completely after test. In such cases, the mass of the specimen with attached wire should be measured before and after the exposure. Mass may be lost during peening if small chips of metal are broken off.

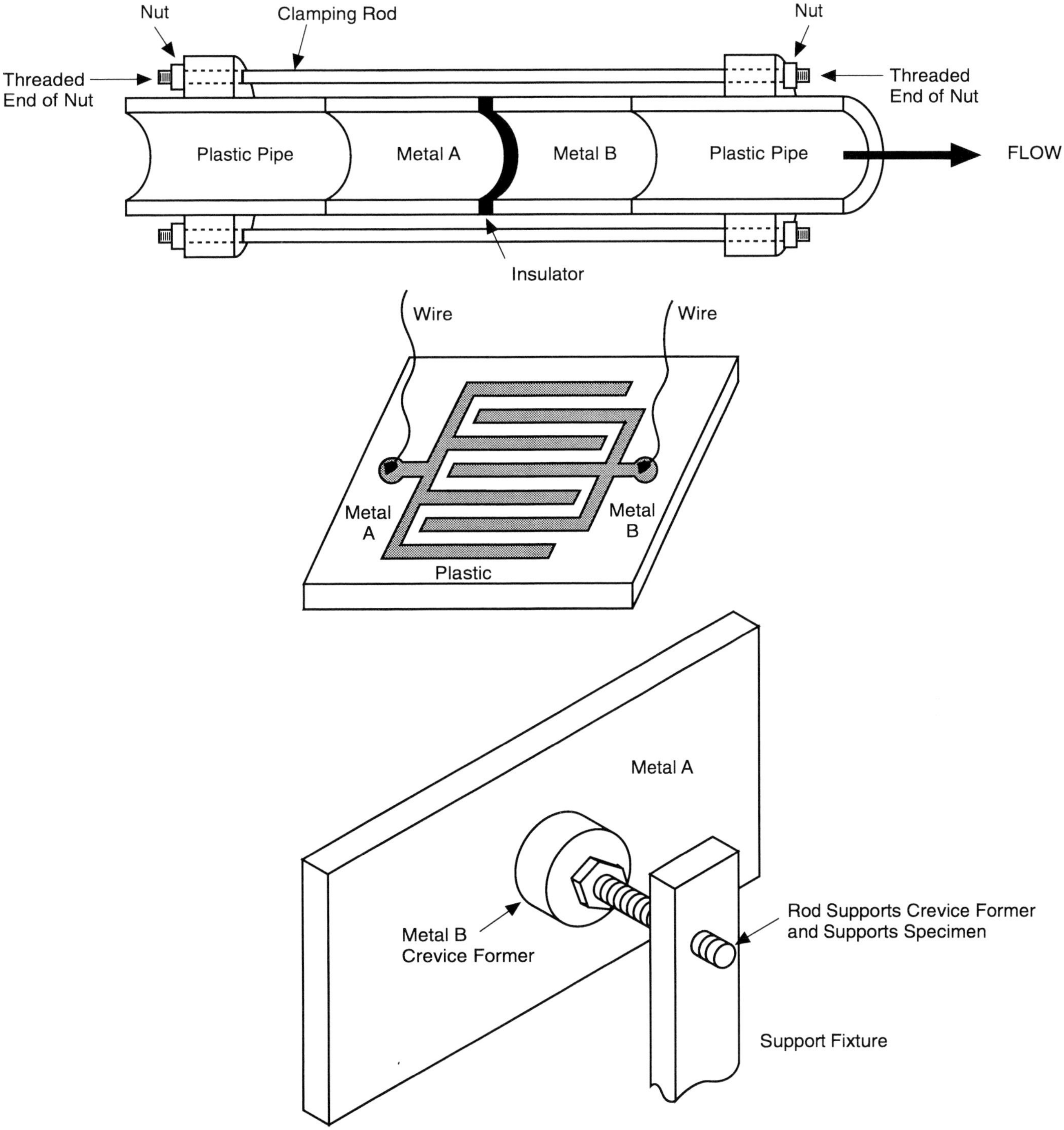

FIGURE 3.11 - Novel mounting methods.

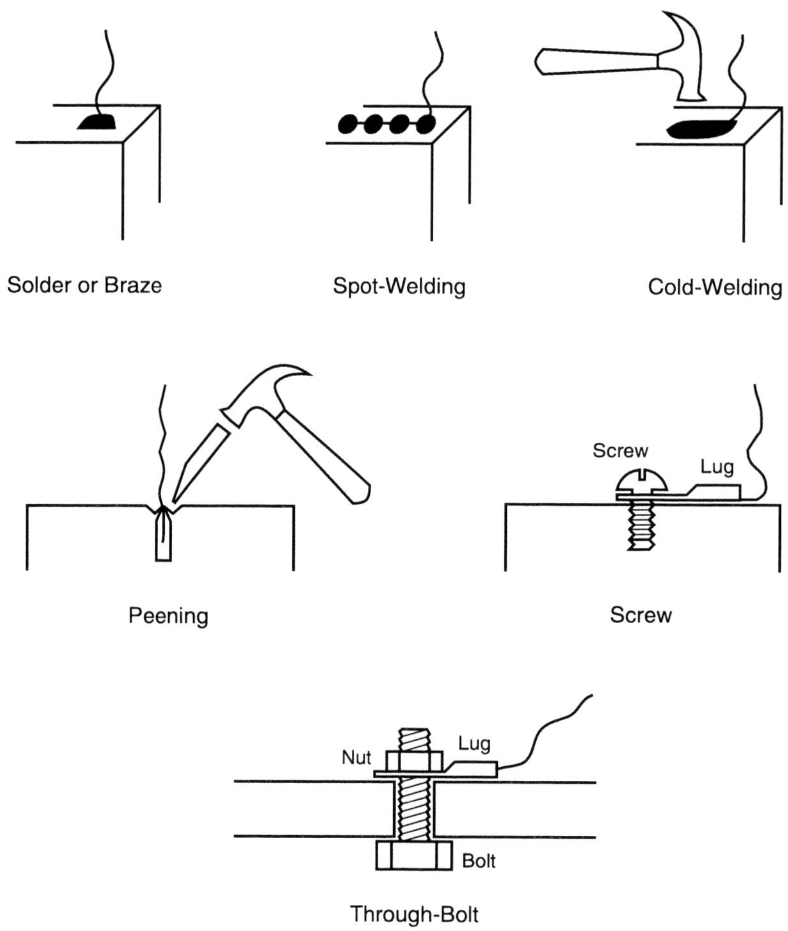

FIGURE 3.12 - Connecting the wire.

Solution Volume

Tests conducted for a considerable period of time or those involving high rates of corrosion may generate a lot of corrosion products that go into the environment and use up components of the environment. For example, corrosion of steel in a mild acid will put ferric ions in the acid and raise the pH. For this reason, long-term exposures should have critical environmental substances periodically monitored if continual environmental replenishment is not used. When the concentration of these substances goes outside limits predetermined by a researcher knowledgeable about the effects of these substances on the material of interest, the environment must be replenished. To prevent frequent replenishment, the volume in the test cell should be large compared to the size of specimens being tested. This is shown in Figure 3.13. For laboratory tests of a few hours where significant corrosion products are not expected, monitoring and replenishment are not necessary provided at least 40 cm^3 of test solution is used for each 1 cm^2 of the combined exposed surface area of the members of the couple. If the intent of the test is to simulate a limited volume of solution, then the volume-to-surface-area ratio should be set the same as in service.

Basic Tests

Galvanic Series

One of the most useful electrochemical measurements for prediction of galvanic corrosion is the corrosion potential of the metals in the uncoupled condition. This

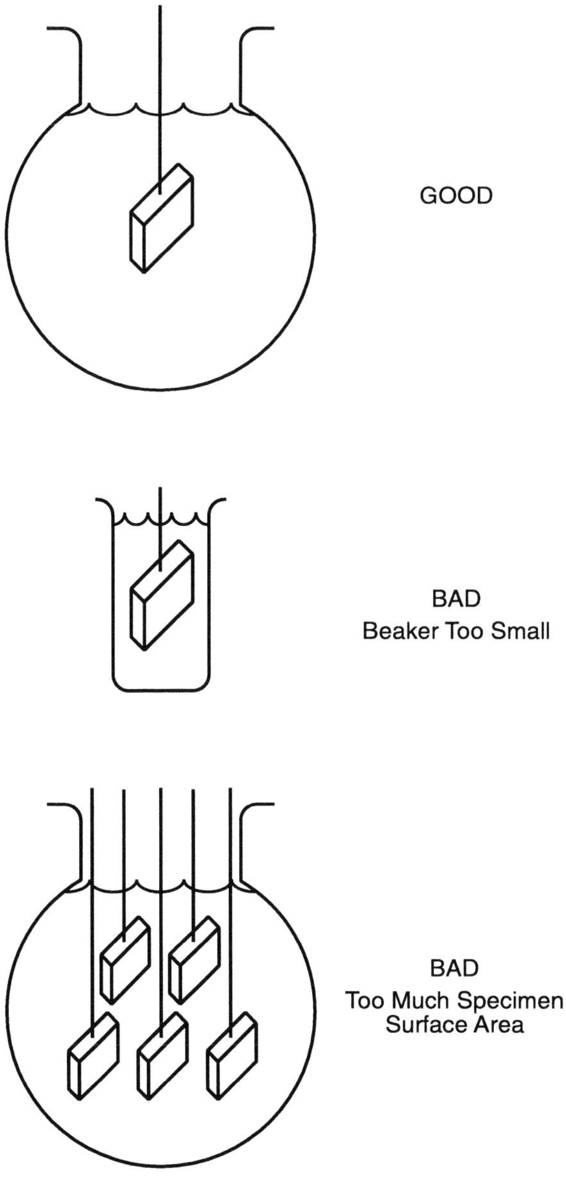

FIGURE 3.13 - Solution volume.

allows prediction of which metal in the galvanic couple could suffer galvanic corrosion. A list of corrosion potentials of a number of alloys in the environment of interest, arranged in order of their potentials, is called a galvanic series. Two methods of presenting galvanic series are shown in Figures 3.14 and 3.15. A material toward the active, or negative, end will tend to corrode faster when coupled to a material located in a positive, or noble, direction relative to it. For example, from Figure 3.14 it can be concluded that mild steel will suffer an increase in corrosion rate in seawater when it is coupled to red brass (the steel is more negative), whereas yellow brass will be unaffected by coupling to red brass (both materials are in the same group). From Figure 3.15 it can be concluded that stainless steel type 302 will corrode faster when coupled to copper in seawater only if the stainless steel is in the active state (dark box) because of a disturbance in the formation of its protective film. If the film is undisturbed, the stainless steel is in the passive state with a very low corrosion rate and the corrosion of the copper will be accelerated.

A galvanic series can predict which metal in a galvanic couple will likely suffer increased corrosion, but it gives no indication of the amount of that increase. The magnitude of the potential difference between two metals bears no relationship to the galvanic corrosion rate experienced by the material with the more negative potential. A guide for development and use of a galvanic series for predicting galvanic corrosion performance is given in ASTM Standard Guide G82.[5] This guide covers not only the use of a galvanic series for galvanic corrosion predictions but also how to make a series, as discussed below.

To make a galvanic series, the environment must first be chosen. If the service environment itself cannot be used, the environment chosen should simulate the environment expected in service as closely as possible. The factors that should be matched are the same as those listed in Chapter 2 and include temperature, flow, and composition. In some cases, for example, in atmospheric corrosion, the exact nature of the electrolyte may not be known, so its composition must be guessed. The accuracy of the test results will depend on the accuracy of the guess. The materials to be tested should also match those of interest, especially in the factors listed in Chapter 2. Finally, an exposure duration should be chosen that is sufficiently long to be indicative of the service condition. If long-term information is desired, the potential must be allowed to stabilize, and thus the exact duration may not be known before the exposure starts.

Panels or components of the materials of interest should be individually mounted and electrical connections made as described earlier. They should then be placed in the environment along with a reference electrode. The reference electrode must be close enough to the specimen or specimens being measured that the resistance of the test solution between the refer-

ACTIVE END
(−)
↑

Magnesium
Magnesium Alloys
Zinc
Galvanized Steel

Aluminum 1100

Aluminum 6053
Alclad

Cadmium

Aluminum 2024 (4.5 Cu, 1.5 Mg, 0.6 Mn)

Mild Steel
Wrought Iron
Cast Iron

13 % Chromium Stainless Steel
Type 410 (Active)
18-8 Stainless Steel
Type 304 (Active)
18-12-3 Stainless Steel
Type 316 (Active)

Lead-Tin Solders
Lead
Tin

Muntz Metal
Manganese Bronze
Naval Brass

Nickel (Active)
76 NI-16 Cr-7 Fe alloy (Active)

60 Ni-30 Mo-6 Fe-1 Mn

Yellow Brass
Admirality Brass
Aluminum Brass
Red Brass
Copper
Silicon Bronze

70:30 Cupro Nickel
G-Bronze
M-Bronze
Silver Solder
Nickel (Passive)
76 Ni-16 Cr-7 Fe
Alloy (Passive)
67 Ni-33 Cu Alloy (Monel)

13 % Chromium Stainless Steel
Type 410 (Passive)
Titanium

18-8 Stainless Steel
Type 304 (Passive)
18-12-3 Stainless Steel
Type 316 (Passive)
Silver

Graphite
Gold
Platinum

↓
(+)
NOBLE or
PASSIVE END

FIGURE 3.14 - Sample galvanic series, list format (From LaQue Center for Corrosion Technology, Inc. Copyright 1966).[6]

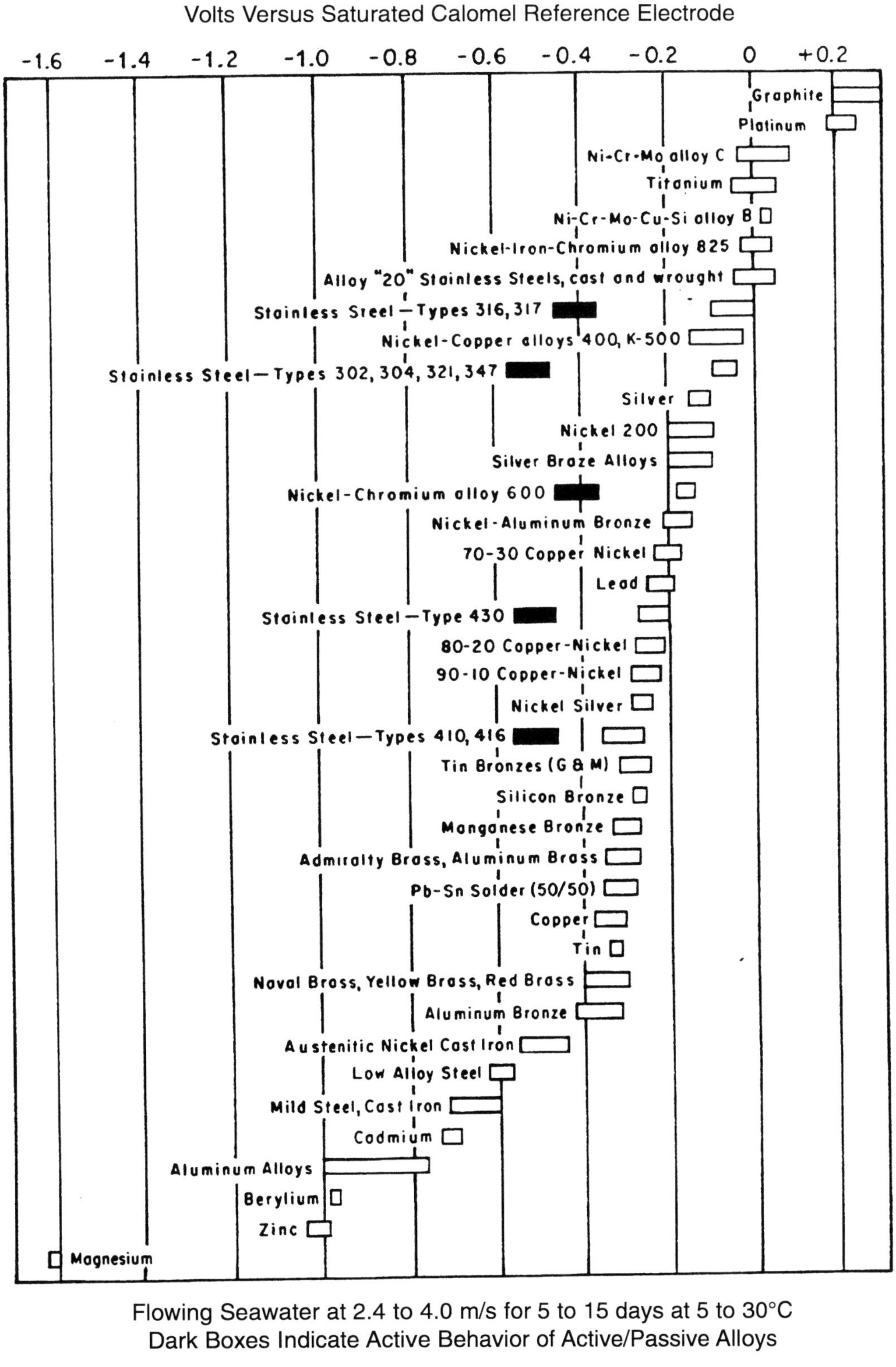

FIGURE 3.15 - Sample galvanic series, numerical format
(Courtesy of LaQue Center for Corrosion Technology, Inc. Copyright 1966).[6]

ence electrode and the specimen is no more than 1/100 of the sum of the resistances of the reference electrode itself and the electrometer measuring circuit. Usually a few centimeters is close enough. The reference electrode should be able to give a stable potential reading over the anticipated duration of the exposure. Several reference electrodes are commercially available. Some common reference electrodes are illustrated in Figure 3.16.

A saturated calomel electrode is typically used in water or acids, as it gives a stable potential with time, with only small changes in potential as temperature varies. In seawater a silver/silver-chloride electrode may be preferable, as it can be used directly in seawater without a glass envelope. In soils, a copper/copper-sulfate electrode is frequently the best choice. The directions for filling and maintaining these cells are supplied by the manufacturer and should be followed carefully. A high-resistance voltmeter, as described below, is required for the measurements. Most reference electrodes cannot pass much current without drifting from the desired reference potential, so a high-resistance voltmeter, such as an electrometer, must be used. The input resistance of the voltmeter (usually marked on the case or meter movement) should ideally be at least 1 teraohm (10^{12} ohms). If metal reference electrodes are used, such as a silver/silver-chloride used directly in seawater, or if the potential measured is only a few millivolts, input resistances as low as 1 gigaohm (10^9 ohms) have been found to be adequate. Measurement error from the voltmeter input resistance can be calculated by first estimating the resistance of the reference electrode. Multiplying this resistance by the potential being measured and dividing by the meter input resistance will give the value of the measurement error as follows (this equation assumes that $R_{reference}$ is much larger than the polarization resistance or the electrolyte resistance):

$$E_{error} = E_{measured} \times R_{reference} / R_{voltmeter}$$

For example, a measured potential of 1V using a reference electrode with a resistance of 10^8 ohms and a volt-ohm meter with an input resistance of 10^9 ohms will result in a potential measurement error of:

$$1 \times 10^8 / 10^9 = 0.1 \text{ V} = 100 \text{ mV}$$

Taking the same measurement with an electrometer with an input resistance of 10^{12} ohms results in a measurement error of:

$$1 \times 10^8 / 10^{12} = 0.0001 \text{ V} = 0.1 \text{ mV}$$

In practice, a measurement is usually taken first with an electrometer, and then with a lower resistance volt-ohm meter. If there is no difference in the two readings and the volt-ohm meter reading is stable, the volt-ohm meter is used for subsequent measurements.

The electrical connections for taking a measurement are shown in Figure 3.17. The common, or negative, terminal on the voltmeter should be connected to the reference electrode, while the positive side is connected to the metal where the potential is to be measured. The potential of each panel as read from the voltmeter should be measured frequently enough to provide an indication of variability as well as systematic drift in the potential.

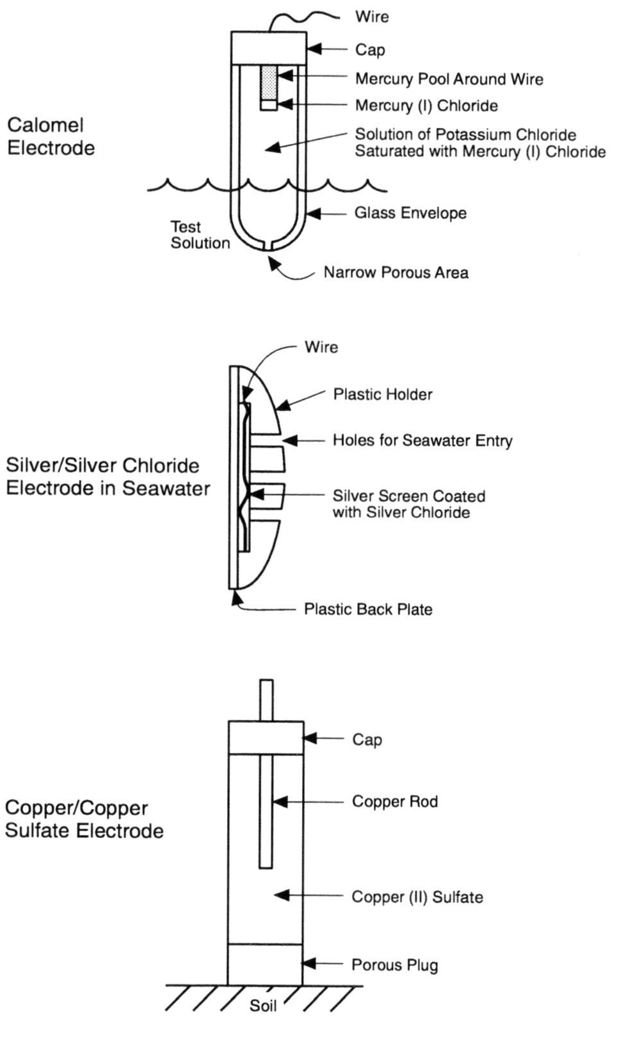

FIGURE 3.16 - Common reference electrodes.

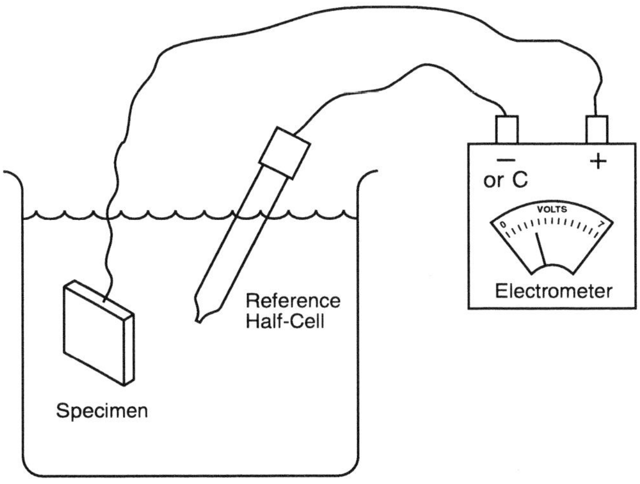

FIGURE 3.17 - Connections for potential measurements.

If both the specimen and the electrometer negative terminal are grounded, the presence of ground loops can lead to measurement errors. This can occur if the specimen is clamped to a grounded test stand and the meter has an internal ground to its line plug's ground pin. Some potentiostats will also ground the specimen to the line plug's ground pin. If ground loops occur, the measured potentials will not be stable or will be much too high. In these cases, the connections to the electrometer can be reversed and the potential recorded opposite in sign to that indicated on the meter, or, preferably, an isolation transformer can be used on the meter to separate its ground from earth ground.

At the conclusion of the test, the data for each panel should be arranged in order of the corrosion potential over the period of interest, as in Figure 3.14. The measured potential for each material may be listed opposite that material as a range, steady-state value, or average with standard deviation. The potential information may even be plotted as a bar graph if desired, as in Figure 3.15. In the case of some materials, two potential ranges may exist, depending on whether the material is in the active or passive state. If two ranges are observed, both should be listed and labeled, the more negative being the active condition.

Galvanic Couple Test

Much more useful information can be obtained if the current of an actual galvanic couple of interest can be measured. This current is related to the rate of metal loss by galvanic corrosion. To measure galvanic currents, specimens of each of the materials in the couple are placed in the environment and electrically coupled together through an external circuit that allows current measurement. In laboratory tests, the resistance of the solution in service sometimes cannot be simulated because of the size or complex geometry of the service components being modeled. In the laboratory, the specimens are usually placed in a relationship to each other that gives the most uniform current distribution so that current density—used for size scaling—can be calculated more easily. To do this, the specimens are typically placed facing each other and a few centimeters apart. The materials and environment are selected following the guidelines in Chapter 2, and specimen mounting and electrical connection are done as described above. The electrical connections to the external circuit may be made through a zero-resistance ammeter or, as illustrated in Figure 3.18, through a small precision resistor (across which a high-resistance voltmeter is placed to measure voltage drop). If a resistor is used, its value should be selected so that the voltage drop across it under the expected current load, calculated as $V = I \times R$, is not more than 5 to 10 mV. Too large a resistor will give a large voltage difference between the two specimens, making this unlike the service condition. Too small a resistor will make the voltage drop too small for accurate current measurements to be made. For specimens 1 to 3 cm on a side in seawater, a good value to start with would be 1 ohm. The resistor may be

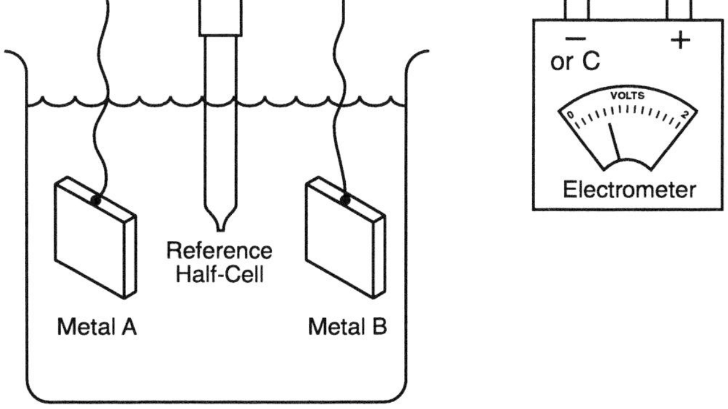

FIGURE 3.18 - Connections for galvanic couple test.

changed during test, providing that the new resistor is connected before the old one is disconnected, thus maintaining electrical contact between the specimens.

A zero-resistance ammeter circuit is shown in Figure 3.19. Commercial instruments typically have two terminals for connection to the specimens, a meter for reading the current, and a sensitivity (range) selection switch. Such a meter can also be constructed from a potentiostat hooked up as in Figure 3.20 and set to 0 mV control potential. Either the resistor or the zero-resistance ammeter may be used for measuring currents flowing in a number of couples, provided that the specimens in each couple are shorted before disconnecting the resistor or meter to move to the next couple.

Current between the specimens should be measured with sufficient frequency to allow determination of variability as well as drift in a single direction. If the total charge passed, calculated as C(coulombs) = I(amperes) [times] time(seconds), is to be calculated to determine the total amount of galvanic corrosion, more frequent current measurements are required, particularly during periods when the current is changing rapidly. Total charge may be most accurately determined from a series of current-vs-time measurements by using the trapezoidal rule as shown in Figure 3.21.

An example of the use of the trapezoidal rule follows:

$$\text{Total Charge} = \frac{.06 + .04}{2}(5 - 0) + \frac{.04 + .03}{2}(15 - 5) + \frac{.03 + .02}{2}(30 - 15)$$

$$= .01 \times 5 + .005 \times 10 + .005 \times 15$$

$$= .05 + .05 + .075$$

$$= .175 \text{ coulombs}$$

Computer software is available to perform this calculation. Alternatively, the total charge passed can be obtained by plotting current vs time, cutting out the graph, and weighing the paper. This weight is compared to the weight of a graph of known total charge, such as 1 mA for 1000s = 1 coulomb, to obtain the total charge passed in the test.

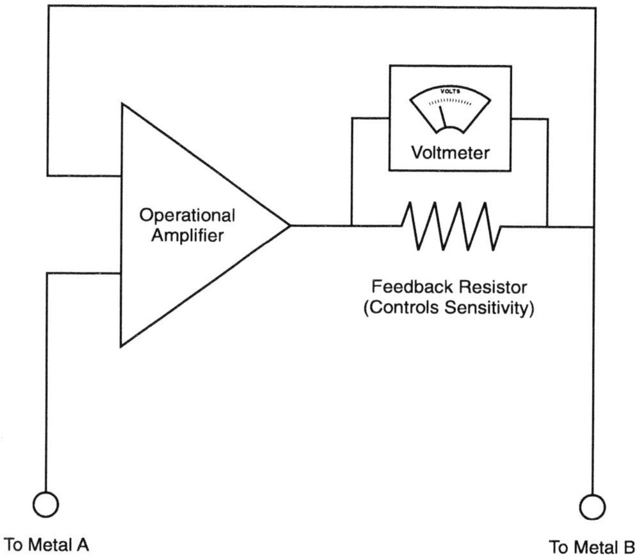

FIGURE 3.19 - Schematic of zero-resistance ammeter.

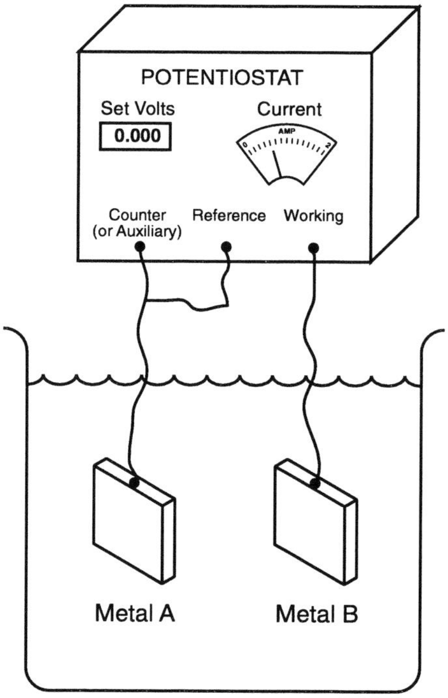

FIGURE 3.20 - Using a potentiostat as a zero-resistance ammeter.

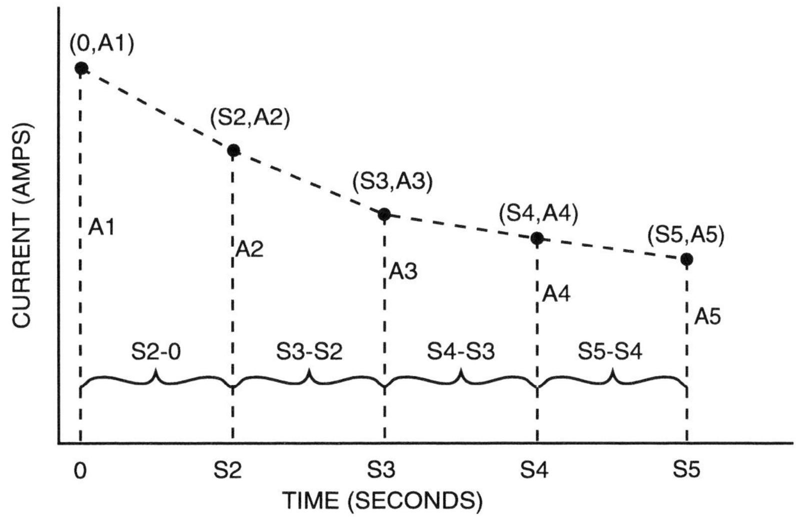

Total Charge (Coulombs) = Total Area =

$$\frac{A1 + A2}{2}(S2\text{-}0) + \frac{A2 + A3}{2}(S3\text{-}S2) + \frac{A3 + A4}{2}(S4\text{-}S3) + \frac{A4 + A5}{2}(S5\text{-}S4)$$

All A's in Amperes

All S's in Seconds

FIGURE 3.21 - Trapezoidal rule for calculating charge passed.

The currents measured and their times are:

Time (seconds)	Current (amperes)
0	0.06
5	0.04
15	0.03
30	0.02

A reference half-cell may be positioned between the two coupled specimens, as in Figure 3.18. The potential of the couple may be measured just as described above under galvanic series, except that the wire from one specimen, while still coupled to the other, is connected to the positive side of the high-resistance voltmeter. Consideration should be given to which terminal of the voltmeter is internally grounded so that ground-loop-induced measurement errors are not introduced, as discussed previously.

Mass-loss measurements on the coupled specimens are extremely useful for determining the amount of galvanic corrosion. For laboratory-sized specimens, a measurement of mass to the nearest ten-thousandth of a gram is usually sufficient for accurate corrosion rate determination. Errors in mass-loss measurement can be caused by incomplete removal of solder or brazing materials, corrosion of the connecting wire, or incomplete removal of corrosion products and scale. Corrosion product removal guidelines and procedures are given in ASTM Standard G1.[7] Other measurements can be taken at the conclusion of the test, including pit or crevice corrosion depth and distribution, corrosion distribution over the surface, metallographic cross-sections to determine depth of selective removal of metal phases (dealloying), and even mechanical properties such as breaking strength.

Additional useful information can be obtained if the anode material and, if possible, the cathode material are exposed to the same environment as the coupled specimens for the same period of time but without being coupled to each other. If this is done, potential and mass-loss measurements should be done on both the coupled and the uncoupled specimens for comparison. This information can then be used to determine the change in corrosion rate caused by the galvanic couple. Usually the anode corrosion rate will increase upon coupling and the cathode corrosion rate will decrease.

Polarization Tests

Galvanostatic Test

Galvanostatic testing is conducted on a single specimen by supplying it with a known constant current and monitoring the resulting potential or mass loss. There is actually little application in galvanic corrosion testing for the galvanostatic test. Its primary use is for determining efficiency—for instance, in ampere-hours per pound—of a material designed to corrode sacrificially to protect another material from corroding. Although sacrificial cathodic protection is a form of galvanic corrosion, a detailed description of this type of test is beyond the scope of this volume. One NACE Standard, TM0190, uses galvanostatic testing to determine efficiency of sacrificial aluminum anode materials.[8]

Potentiostatic Test

Potentiostatic testing is conducted on a single specimen by holding it at a known constant potential relative to a reference electrode and monitoring the current required and mass lost. When a series of tests is conducted over a range of different potentials, the potentiostatic test is an extremely valuable tool, providing one of the most accurate quantitative predictions of galvanic corrosion behavior in service.

The galvanic corrosion behavior of a component in service can be predicted by a superposition of polarization curves, as described in Chapter 1. The polarization behavior of many galvanic systems is time dependent however, because of the formation of corrosion product films, roughening of the surface, deposition of material from solution on the surface, selective corrosion of certain components from the alloy, and other factors. This leads to time-dependent galvanic corrosion in service. For accurate prediction, it is therefore necessary to use a polarization curve that is developed over a time frame similar to the service life—or at least long enough to ensure stability of the curves. The potential of a component in service often is close to being constant. Optimum accuracy is therefore achieved if the curves are also developed under constant potential conditions. Curves developed in this way are called long-term potentiostatic polarization curves.

Long-term potentiostatic polarization curves are generated by running a series of potentiostatic tests on the materials of interest, as described in detail in the volume on Electrochemical Testing in this Corrosion Testing Made Easy series. Figure 3.22 shows the electrical connections for a single potentiostatic test. For each test at a given potential, current is monitored over time until stability is reached. The value of the stable current for this potential is recorded. The test is then repeated at other potentials, yielding a series of potential/current-data pairs. Each pair defines a point on a polarization diagram. When enough points are generated, a connect-the-dots procedure can be followed to generate the curve. This procedure is illustrated in the left half of Figure 3.23.

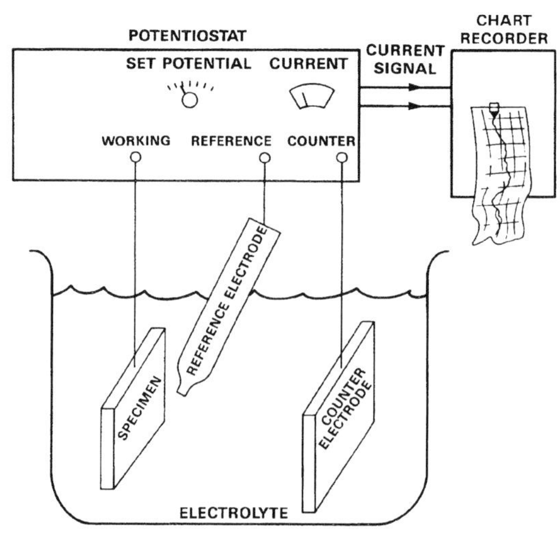

FIGURE 3.22 - Electrical connections for potentiostatic test.

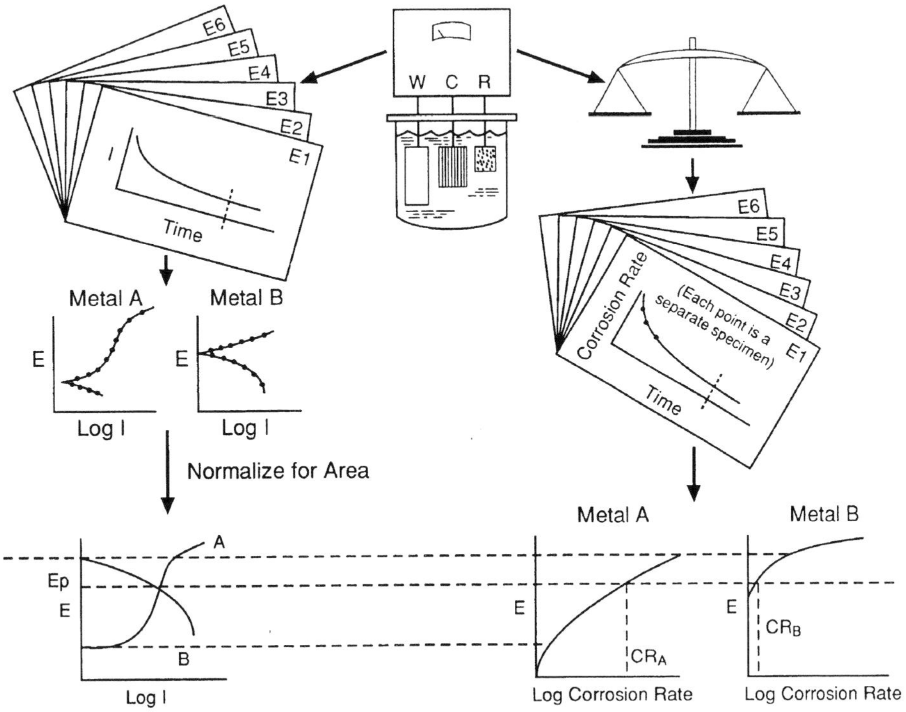

FIGURE 3.23 - Developing a polarization curve from potentiostatic tests.

An added bonus can be obtained by measuring the mass loss on each specimen tested. This will yield a series of potential/mass-loss data pairs, which can be used to obtain a plot of corrosion rate as a function of potential. This procedure is shown on the right side of Figure 3.23. The superposition of the long-term potentiostatic polarization curves for each material can be used to predict a coupled potential. This coupled potential can be used in conjunction with the corrosion rate curves to predict corrosion rate, as shown in the bottom of Figure 3.23. This avoids the inaccuracy of predicting corrosion rate from galvanic current using Faraday's Law.

Potentiodynamic Test

The above technique for generating long-term potentiostatic polarization curves involves a large number of specimens and potentiostats—and a long period of time. This is frequently too costly to be done in practice. A simpler, faster, and cheaper—although less accurate—way to generate polarization curves is by the use of a potentiodynamic polarization test. In this test, a single specimen is used for each material. The potential of that specimen is controlled while monitoring current, but the value of the controlled potential is slowly swept over the range of interest. Slow potential scan rates should be used to avoid errors caused by charging of the electrochemical interface and to allow corrosion products or precipitates to form and become stable. Usually these slow scans are conducted no faster than 10 mV per minute. When little time-dependent polarization behavior is expected, as in situations where the surfaces remain clear of corrosion products and precipitates, faster scan rates can be used.

The electrical connections for a potentiodynamic polarization test are dependent on the instrumentation used, and the instructions from the manufacturer of the potentiostat used should be consulted. A typical configuration is shown in Figure 3.24. Because mass-loss data are not relevant in this type of test, corrosion rate information must be deduced from the galvanic current predicted by superposition of the curves. This is done by the use of Faraday's Law, as described in Chapter 1. The typical inaccuracies in the use of Faraday's Law also apply here. In addition, if the coupled potential of the anode lies within roughly 50 mV or less of its corrosion potential, the magnitude of corrosion may be se-

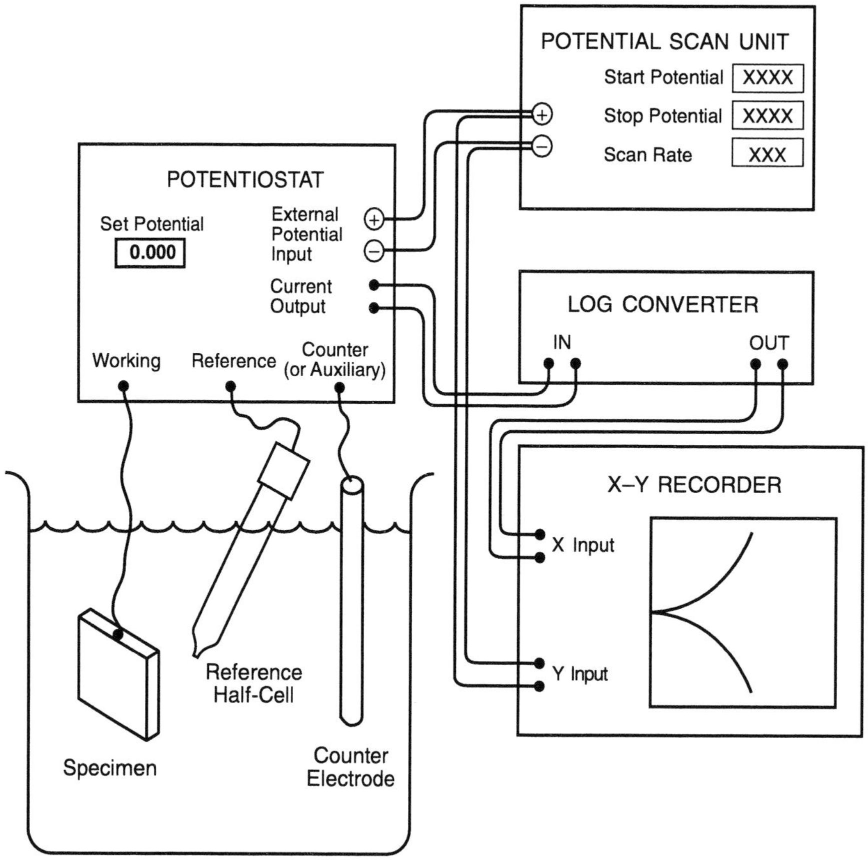

FIGURE 3.24 - Connections for potentiodynamic test.

verely underestimated, as shown in Figure 3.25. A better estimate, in this case, is to add the rate calculated from Faraday's Law to the rate of corrosion of the freely corroding anode material. Usually this will slightly overestimate the actual corrosion rate of the anode material in the couple, a much more conservative procedure.

Pre-exposure

In cases where the polarization behavior of the alloys of interest is significantly affected by time and where the alloys' corrosion potentials are within 60 mV, the accuracy of a potentiodynamic test may be improved by pre-exposure of each specimen before applying any external control on the potential. During pre-exposure, the corrosion potential of the specimen should be monitored as one indicator of when stability in the corrosion reactions has been reached. Another indication of stability could be the formation of corrosion products that do not change in appearance with time. After stability is reached, or when what is deemed to be a reasonable period of time has elapsed, the potentiodynamic test is started.

For alloys not more than 60 mV different in potential that are to form a galvanic couple in service, pre-exposure will be at a potential close to that experienced by the material in the couple. This approximates the potentiostatic test described above and should provide more accurate polarization curves than a potentiodynamic test without pre-exposure. Unfortunately, most galvanic couples of interest will consist of materials more than 60 mV different in potential, because those similar in potential will likely experience little galvanic corrosion. In such systems, the improvement in accuracy obtained by pre-exposure is not great, and the test time is significantly extended.

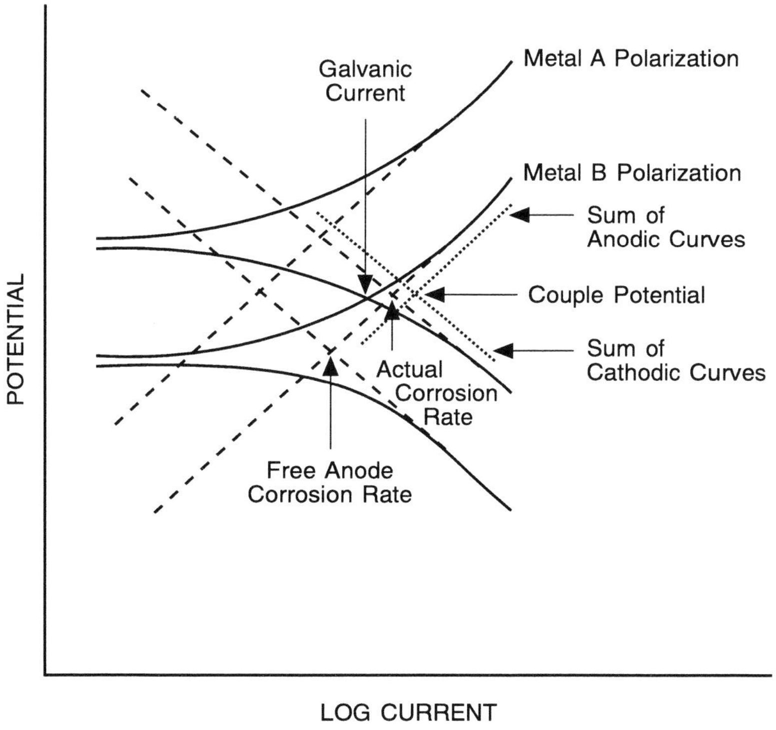

FIGURE 3.25 - Reason for inaccurate prediction of corrosion rate in galvanic couple where potentials are close.

References

1. ASTM G30-90, "Standard Practice for Making and Using U-Bend Stress-Corrosion Test Specimens." (Philadelphia, PA: ASTM, 1990).

2. ASTM E8-93, "Standard Test Methods for Tension Testing of Metallic Materials." (Philadelphia, PA: ASTM, 1993).

3. ASTM G71-81, "Standard Guide for Conducting and Evaluating Galvanic Corrosion Tests in Electrolytes." (Philadelphia, PA: ASTM, 1981).

4. ASTM G82-83, "Standard Guide for Development and Use of a Galvanic Series for Predicting Galvanic Corrosion Performance." (Philadelphia, PA: ASTM, 1983).

5. EG&G Instruments – Princeton Applied Research (Princeton, NJ: 1993).

6. LaQue Center for Corrosion Technology, Inc. (Wrightsville Beach, NC: 1966).

7. ASTM G1-90, "Standard Practice for Preparing, Cleaning, and Evaluating Corrosion Test Specimens." (Philadelphia, PA: ASTM 1990).

8. NACE TM0190-90, "Impressed Current Test Method for Laboratory Testing of Aluminum Anodes." (Houston, TX: NACE, 1990).

Bibliography

General

Arnold, C.G. "Corrosion Testing Techniques." *Corrosion and Corrosion Protection Handbook.* Ed. P.A. Schweitzer. New York, NY: Marcel Dekker, Inc., 1983.

Baboian, R., L. McBride, R. Langlais, G. Haynes. "Effect of Modern Electronics on Corrosion Technology." *Materials Performance* 18, 12 (1979): pp. 40-44.

Champion, F.A. *Corrosion Testing Procedures.* London,

UK: Chapman and Hall, 1952.

Fontana, M.G. *Corrosion Engineering*. New York, NY: McGraw-Hill, 1986.

Hack, H.P. "Evaluation of Galvanic Corrosion." *Corrosion*. Metals Handbook, Vol. 13. Materials Park, Ohio: ASM International, 1987.

Rowe, L. C., R.L. Chance. "Fundamentals of Corrosion Testing." *Automotive Corrosion by Deicing Salts*. Ed. R. Baboian. Houston, TX: NACE, 1981.

Shreir, L.L., ed. "Corrosion Control." *Corrosion*, Vol. 2. London, UK: Newnes-Butterworth, 1976.

Polarization Testing

Baboian, R. "Prediction of Galvanic Corrosion Using Electrochemical Techniques." *Electrochemical Techniques for Corrosion Engineering*. Ed. R. Baboian. Houston, TX: NACE, 1986.

Dean, S., Jr. "Electrochemical Methods of Corrosion Testing." *Electrochemical Techniques for Corrosion Engineering*. Ed. R. Baboian. Houston, TX: NACE, 1986.

Hack, H.P. "The Potentiostatic Technique for Corrosion Studies." *Electrochemical Techniques for Corrosion Engineering*. Ed. R. Baboian. Houston, TX: NACE, 1986.

Scully, J.R. "Electrochemical Methods of Corrosion Testing." *Corrosion*. Metals Handbook, Vol. 13. Materials Park, Ohio: ASM International, 1987.

Siebert, O.W. "Applications of the Rapid-Scan Potentiodynamic Polarization Technique for Corrosion Studies." *Electrochemical Techniques for Corrosion Engineering*. Ed. R. Baboian. Houston, TX: NACE, 1986.

CHAPTER 4

Component Testing and Scale Modeling

Full-Scale Components

One way to determine if a particular set of materials will cause galvanic corrosion in a given piece of hardware is to build that piece of hardware, expose it to the environment in which it will operate, and watch to see what happens. Sometimes it may be desirable to expose full-scale components in an environment different from that in which they will operate—to reduce expense, increase corrosion rate, or simulate the times when the environment changes (for example, during a toxic waste spill). For the purposes of the following discussion, full-scale testing will refer to any testing involving full-size hardware regardless of the environment. Full-scale component testing is an extremely useful galvanic corrosion test tool, but it is also frequently misapplied or prohibitively expensive. The following are descriptions of considerations regarding selection and application of full-scale testing. A summary of the advantages and disadvantages of full-scale testing is included as Table 4.1.

Table 4.1
Full-Scale Component Testing

Advantages	Disadvantages
Easy to understand	Can be expensive
Easy interpretation of results	Operating conditions may not be reproducible
May be able to get more accurate measurements in lab than in service	Long exposure periods
Geometric factors are reproduced exactly	Some types of damage hard to find before failure
	Performance may depend on presence of other structures

Advantages

Full-scale component testing is the easiest of the galvanic corrosion testing methods to understand. The test specimen is the structure of interest made of the materials of interest. For example, if pump corrosion is of concern, one simply obtains an extra pump, runs it in the environment it would see in service, and observes the results. No great thought need be given to electrochemistry, area ratio effects, geometrical effects, modeling, or other more theoretical considerations. Likewise, interpretation of the results is straightforward.

Corrosion or pitting rates need not be translated into service performance. Either the component works at the conclusion of the exposure or it does not. It may be prohibitively expensive to remove an actual working component from service and transport it to a laboratory for accurate measurement of the degree of corrosion, or it may be too difficult to make controlled measurements in a service situation. Therefore, more accurate evaluations of corrosion performance are sometimes possible if the exposures are performed on full-scale components in the laboratory. If the corrosion rate stabilizes quickly, shorter exposure periods can be used than in service, because a smaller amount of corrosion can be detected using standard laboratory techniques. At first glance this appears to be the ideal way for predicting galvanic corrosion performance, particularly where the component is inexpensive to manufacture or where geometrical and modeling effects are difficult to predict.

Disadvantages

Some components are too expensive to construct just for a test. For example, an offshore oil platform would be a multibillion-dollar test specimen. In other cases, the operating conditions of the component are difficult to reproduce in a test. For example, a proper test of a heat exchanger may require an entire plant to provide the proper input conditions of heat transfer, flow, and water chemistry. Extrapolating to a long service life the condition of a component after a short period of exposure can be difficult or impossible because it is hard to determine when the corrosion rate has stabilized. Full-scale testing relies on accurate corrosion assessment methods; however, the most sensitive corrosion measurement techniques—electrochemical methods and accurate mass-loss measurement—cannot usually be used on massive components. This means that useful answers usually require the component in question to be tested for a long period compared to laboratory testing. In some cases, the galvanic corrosion damage is difficult or impossible to assess before failure occurs. For example, if galvanic corrosion creates hydrogen damage, a catastrophic failure may have the potential to occur in a period of time longer than the test with little way to measure cumulative damage earlier. This is particularly true of corrosion fatigue damage of components under repetitive varying load. Some components cannot be isolated from the surrounding structures without creating conditions different from those in service. For example, it would be useless to test a ship rudder assembly by itself if in service it would be mounted on a cathodically protected ship hull.

Conducting Full-Scale Component Tests

Geometry

The geometry of the component being tested must be as close as possible to the geometry of the one used in service. This includes the shape of the component and the pieces of which it is constructed and its relationship to other components in the structure. For example, if a component is to be operated inside a steel tank and electrically connected to the steel, then in the test setup, that component must be connected to a similar area of steel. Similarly, a piping valve under test must be connected to an appropriate area of piping material. If flow is a factor, the inlet and outlet sections should be similar to those to be connected to the component in service. If the component will be in an electrical field created by cathodic protection or another galvanic couple, then the field must be modeled as well.

Initial Conditions — History of Surface

It is sometimes difficult to get the surface condition of the materials in a specially constructed test component to be similar to those of a mass-produced service component. Installation of a component in an uncontrolled service environment may lead to damage not anticipated in a test component. For example, the surface of components installed in a service environment may be damaged by dropping the component, dropping tools on the component, or marking the component's surfaces with paint, dyes, or adhesive tape.

Operating Cycle

Possibly one of the most difficult components to test is a large pump. The appropriate environment must be supplied. This usually means using a recirculating system to conserve the test environment. Heat build-up and accumulation of corrosion products, or depletion of important chemicals in the environment may become a problem during operation of such test equipment. In service, pumps may not see continuous duty; they may experience periods of shutdown. Even if continuous operation is planned, breakdowns can occur. In service, a period of no flow may also occur after installation and before start-up of the plant. Hydro-testing (pressure testing with water) may produce an initial environment quite different from the later operating environment. Establishing a realistic operating cycle for the component in test can, therefore, be quite difficult.

Measurements

The measurements that can be taken to evaluate the corrosion performance of a large, complicated component are summarized in Table 4.2. These measurements are much more limited than those that can be taken on specimens in a laboratory. Galvanic current measurements are usually not taken because the dissimilar metals are usually in intimate electrical contact. In some cases, a specially constructed component can be prepared with the dissimilar metals electrically isolated and

Table 4.2
Measurements to Evaluate Performance of Complex Component

Galvanic current—on specially constructed components only
Coupled potential—dependent on reference cell location
Mass loss—on separable small components only
Visual observations
Dye penetrant or magnetic particle
Depth of pitting
Depth of attack—limited situations only
Destructive evaluation—including metallography

connected externally through a current-measuring circuit. For example, a stainless steel wire rope to be used on a steel drilling platform may be electrically isolated from the platform with nonmetallic bushings and then connected back to the platform through an ammeter to monitor the galvanic current. The coupled potential of the test structure can sometimes be obtained by measurement vs a suitable reference electrode, but in cases where current flow and environmental resistivity are both large, this potential reading will vary with the location of the reference electrode.

In some cases, mass loss of the materials in the component may be measurable, but frequently the pieces cannot be disassembled or are too heavy for accurate mass-loss measurement. Usually, post-test evaluation is restricted to visual observations of the amount and distribution of corrosion, special techniques for inspection for cracks (such as the dye-penetrant or magnetic-particle methods), pit-depth measurements using a micrometer gauge, and, if corroded and uncorroded material are close together, evaluation of the depth of uniform attack by measuring the step height between corroded and uncorroded surfaces. In some cases pre- and post-test dimensional measurements can be compared to find attack depth. Destructive evaluation of corroded pieces, including the use of metallography, may also be helpful in post-test evaluation.

Scale Modeling

At first glance it would seem that a great deal of the expense of full-scale component exposures could be reduced, with little sacrifice, by the use of a scale model of the component. Scale modeling can introduce some inaccuracies, however, particularly in cases where nonuniform potentials or currents are expected. Some examples of common situations where the geometry causes nonuniform distribution of potentials and currents are the internal surfaces of tubes, grates, screens, and internal corners of tanks.

Scale modeling in galvanic corrosion tests can be conducted by simple size scaling or by scaled conductivity tests, described below. It is necessary to understand scaling laws to appreciate the difference between the two scaling methods.

Scaling Laws

Galvanic current is not uniformly distributed over the surfaces of the coupled materials because of the electrical resistance of the solution. This resistance makes it easier for the ions to move short distances to nearby structures than to move large distances to distant structures. Also, when ions flow through a restricted space such as a tube, some are consumed by reactions on the side walls, making it more difficult for them to make it to the tube end. This is illustrated in Figure 4.1 (see Page 44), which shows zinc cathodic protection delivered to the inside of a tube from one end.

The galvanic current distribution over a structure is controlled by the resistance to current flow at various points on the structure. This resistance at any point is the sum of several components, as follows:

$$R_T = R_S + R_{PC} + R_{PA} + R_M$$

where: R_T = total resistance to current flow

R_S = resistance through the Solution

R_{PC} & R_{PA} = resistances to Polarization of the Cathode and Anode materials, respectively

R_M = Metallic path resistance (including bulk resistance of anode and cathode materials and resistance of any electrical connection between them)

R_M can usually be measured or calculated from bulk resistivity and the geometry, but it is usually negligible compared to the other terms. The voltage drops introduced by a galvanic current flowing through these re-

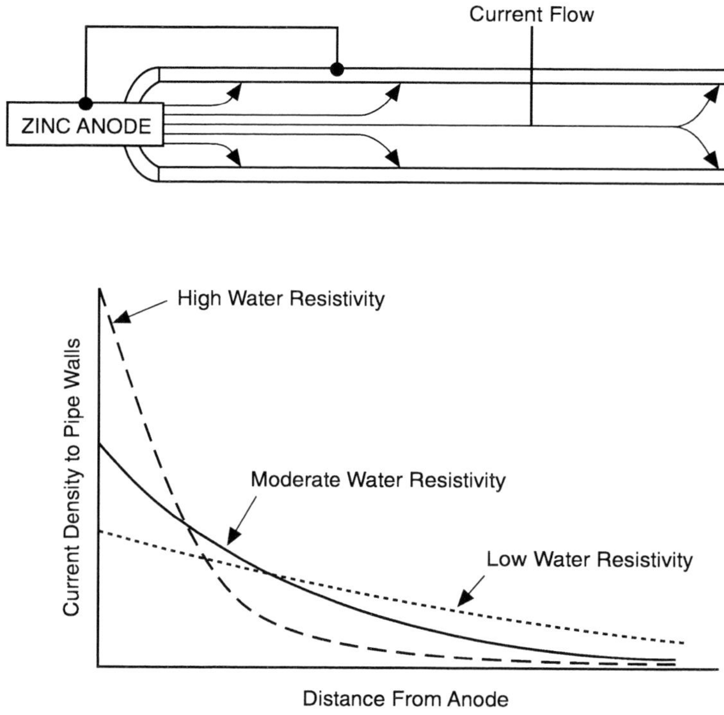

FIGURE 4.1 - Galvanic current along the inside of a pipe.

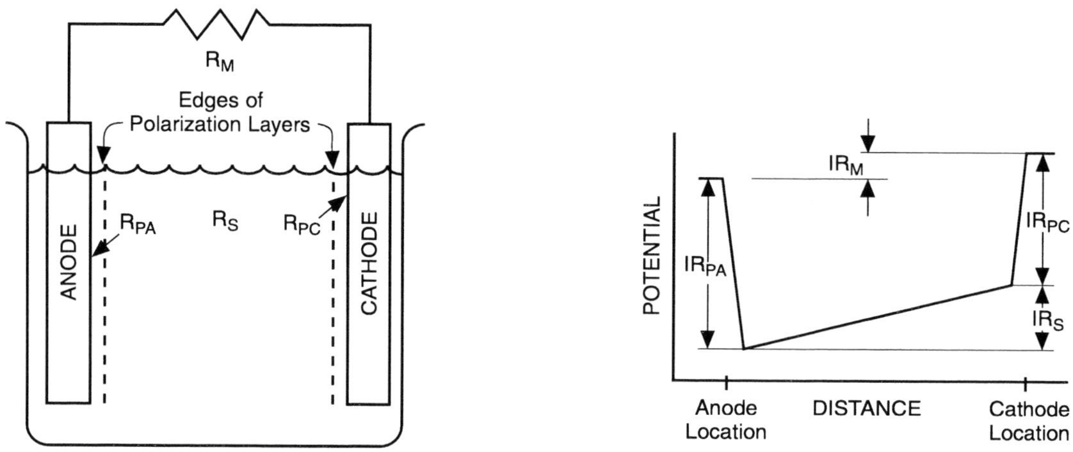

FIGURE 4.2 - Potential variation across a galvanic cell.

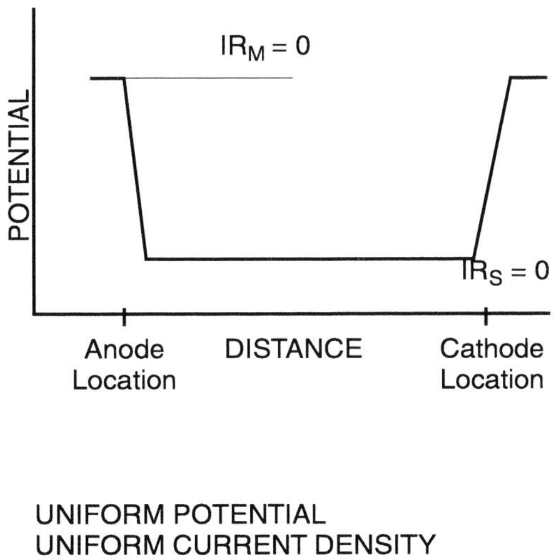

FIGURE 4.3 - Potential variation when metallic path resistance, R_M, and solution resistance, R_S, equal zero.

sistances as one goes from the anode to the cathode in a simple galvanic cell are illustrated in Figure 4.2.

Simple Size Scaling

Simple size scaling may be done just by changing the size of all parts of the component by an equal factor. This works only if R_S is negligible, as in seawater or strong acids over small distances, making current distribution uniform over the surfaces. This situation is illustrated in Figure 4.3.

Scaled Conductivity Testing

Scaled conductivity modeling is based on the following: The linear dimensions of all parts of the component are reduced by some scale factor F. For example, assume that a 1/10 scale model is desired, as seen in Figure 4.4. All linear dimensions will be scaled by a factor of 10. This has the effect of reducing the areas of all surfaces in our example by a factor of 100. Current flows through the solution without crossing its own path. If a tube-shaped volume of solution is isolated between the anode and the cathode, as in Figure 4.5 (see Page 46), the resistance of this tube of solution can be calculated as $R_S = \rho$ [times] L / A, where ρ is the solution resistiv-

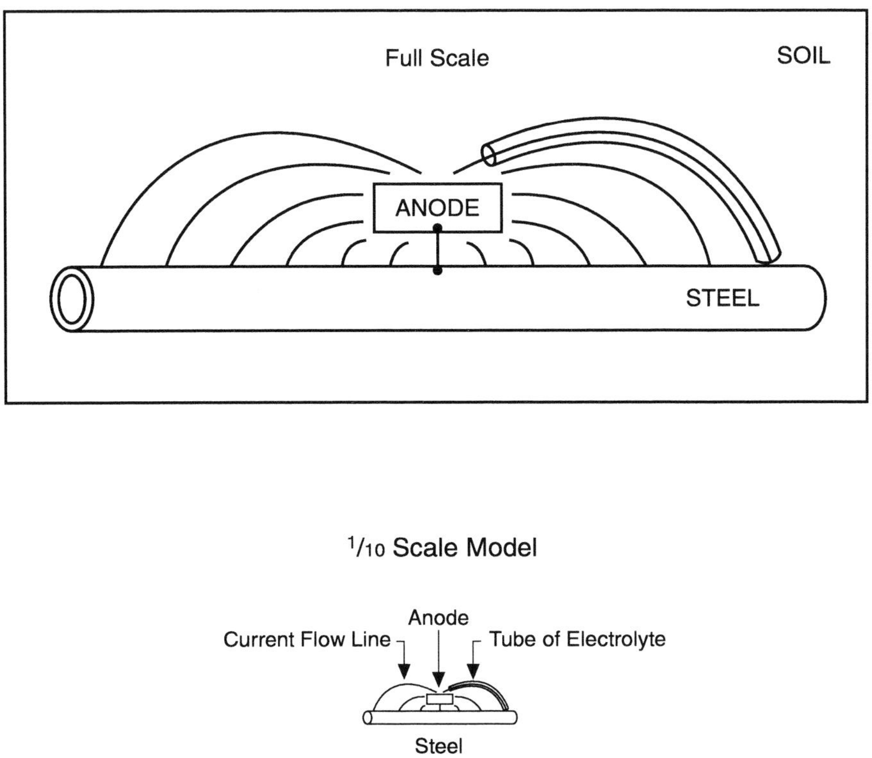

FIGURE 4.4 - Tube of electrolyte for modeling.

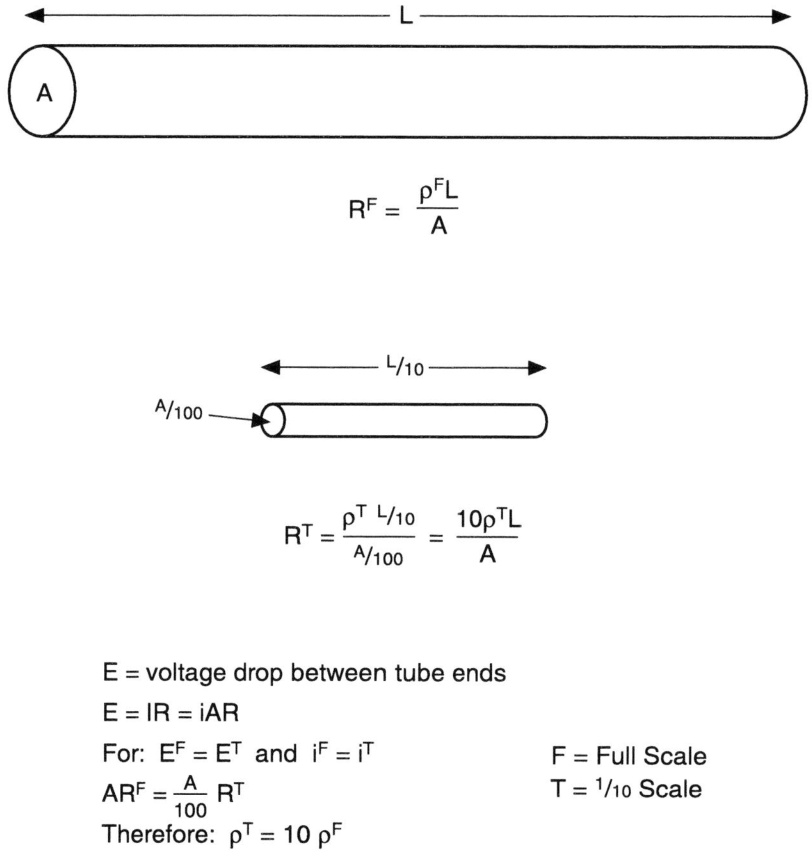

FIGURE 4.5 - Electrolyte tubes in 1/10 scale model.

ity, L is the length of the tube, and A is its cross-sectional area.

In a 1/10 scale model, R_S is increased by a factor of 10. To maintain the same current densities on the scaled area, which is smaller by a factor of 100, the solution conductivity must be decreased by a factor of 10. In general, if the model is reduced by a factor of F, then the conductivity of the solution must be reduced by this same factor of F. This is called conductivity scaling. In some aqueous systems such as seawater or acids, this can be done by dilution with distilled water.

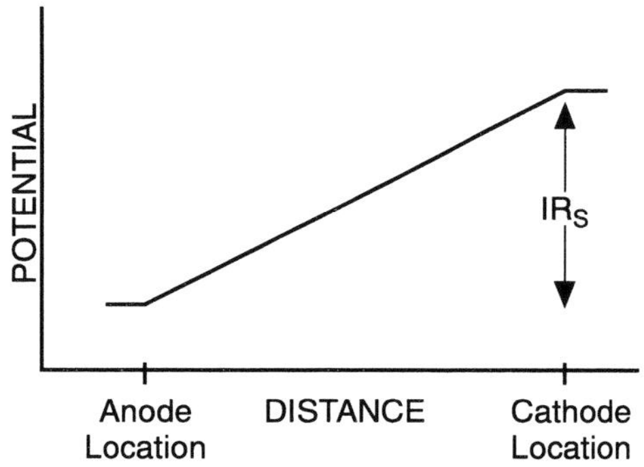

FIGURE 4.6 - Potential variation when R_{PC} = 0 and R_{PA} = 0.

Conductivity scaling works only if R_{PC} and R_{PA} are negligible with respect to R_S, as illustrated in Figure 4.6. This is because changing the conductivity of the solution will likely change the value of R_{PC} and R_{PA}. Only if these terms are negligible compared to R_S will conductivity scaling result in accurate modeling, although testing of this type has been done with apparent success even if these conditions are not met.

Selecting the Type of Scaling

If the solution resistance is small, as in strong acids over short distances, potentials and currents will be uniformly distributed on the component. In this case, scale modeling can be performed without conductivity scaling. If the solution resistance is high, as in river water, and the corrosion rates of the materials in the couple are high, then the R_P terms are small and conductivity scaling may be attempted. This situation can be ascertained only by polarization resistance measurements and approximate solution resistance calculations. In practice, conductivity scaling is seldom used because ideal conditions for its use are rarely encountered.

If neither the R_S nor the R_P terms are negligible, then scale modeling will not give accurate results and should not be attempted. Computer modeling is then the only method to use to obtain quantitative results.

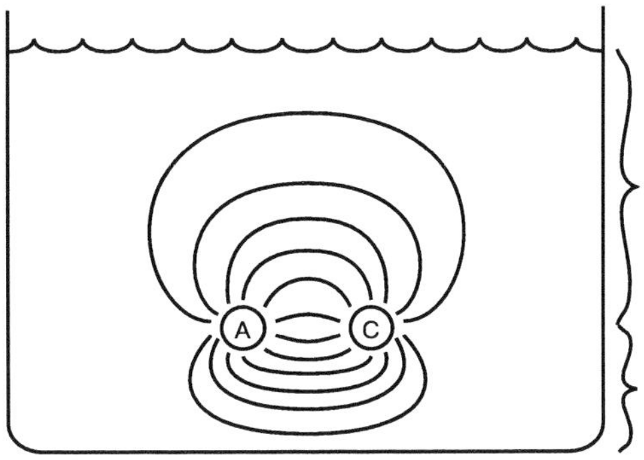

Figure 4.7 - Current constriction in small tank.

Tank Size — Wall Effects

A scale model is usually placed in a tank for exposure. If the tank walls are close to the model, the current flow path through the solution around the model will have a reduced cross section, resulting in larger voltage gradients in the solution than if the walls were not present. This can be seen in Figure 4.7. A rule of thumb to avoid wall effects is to keep all tank walls at least as far away from the model as five times the largest dimension of the model. Tank walls should always be nonconductors, to prevent current from taking the low-resistance path through a conductive tank wall rather than the high-resistance path through the solution. If walls or boundaries occur in the full-scale structure in service (such as the surface of the water or the bottom of a lake), the tank walls will need to model the boundaries of this environment.

Measurements

Measurements to take on a scale model are similar to those on a full-scale component, as shown in Table 4.2. These include mass loss, visual assessment of corrosion damage, pit-depth measurements, inspection for cracks, dimensional changes, galvanic currents if possible, and destructive evaluations. Measuring the potential of the model is useful when the conductivity has not been scaled.

In scaled conductivity tests where all the previously discussed caveats are satisfied, the potential distribution over and around the scaled component should be similar to that of the full-scale component. In this case, potential measurements at various points on and around the component may also prove valuable. These measurements are taken by connecting a voltmeter between the component and a reference electrode. Care must be taken that the wire to the reference electrode be totally insulated from the solution through which it must run, as even a pinhole in the insulation will result in erroneous measurements. The electrode is placed at a location of interest in the solution near the component and the voltage is noted. The electrode is then moved to a new location and the measurement repeated. Enough locations are selected to give a representation of the distribution of potential around the component. Where the potential changes rapidly over a small distance, readings should be taken closer together.

Choosing Full-Scale or Scale-Model Tests

Full-scale component testing should be used whenever cost permits. If full-scale components are too costly and if the resistance through the solution, R_S, is much greater or much less than the polarization resistances, R_{PA} and R_{PC}, then the use of scale modeling is appropriate. In analyzing the results of a scale-model study, keep in mind that scale models may not accurately portray actual galvanic corrosion. In addition, they may not properly simulate stresses, crevice geometries, coated surfaces, or flow conditions.

Bibliography

General

Baloun, C.H. “Corrosion Testing in Water.” *Corrosion.* Metals Handbook, Vol. 13. Materials Park, Ohio: ASM International, 1987.

Bimetallic Corrosion. Publication No. 14. National Corrosion Service, Department of Industry, England: Colibri Press Ltd., 1982.

Champion, F.A. *Corrosion Testing Procedures*. London, UK: Chapman and Hall, 1952.

Escalante, E. "Corrosion Testing in Soil." *Corrosion*. Metals Handbook, Vol. 13. Materials Park, Ohio: ASM International, 1987.

Money, K.L. "Corrosion Testing in the Atmosphere." *Corrosion*. Metals Handbook, Vol. 13. Materials Park, Ohio: ASM International, 1987.

Rowe, L.C., R.L. Chance. "Fundamentals of Corrosion Testing." *Automotive Corrosion by Deicing Salts*. Ed. R. Baboian. Houston, TX: NACE, 1981.

Uhlig, H.H., ed. *Corrosion Handbook*. New York, NY: John Wiley and Sons, 1948.

Scale Modeling

Clerbois, L.E., F. Heintz, F. IJselling, J. Rowlands, J. Simpson. "Principles for Scaling of Corrosion Tests." *British Corrosion Journal* 20, 3 (1985): pp. 107-116.

DeCarlo, E. "Computer Techniques for Offshore Corrosion Protection." *Sea Technology*, September 1981.

Hack, H.P. "Scale Modeling for Corrosion Studies." *Corrosion* 45, 7 (1989): pp. 601-606.

Tighe-Ford, D.J., J.N. McGrath, L. Hodgkiss. "Design Improvements for a Ship's Cathodic Protection System Using Dimensions and Conductivity Scaling (DACS)." *Corrosion Prevention and Control* 32, 5 (1985): pp. 89-91.

CHAPTER 5

Atmospheric Testing

Special Circumstances

Galvanic corrosion testing in the atmosphere can be fairly easily standardized because of special circumstances—discussed below—that limit the number of variables. Three atmospheric tests will be described in this chapter.

Three factors that are different in atmospheric testing than in most other galvanic corrosion testing are (1) the presence of a limited amount of electrolyte, (2) the need for a very long duration of exposure, and (3) the importance of specimen orientation.

Limited Electrolyte

The thin, relatively nonconductive water layer that forms on atmospheric specimens limits the galvanic interaction distance to approximately 10 mm. Thus, area ratio effects are usually not significant. The small amount of electrolyte also prevents the use of conventional electrochemical techniques, including electrochemical monitoring. Galvanic current measurements require that the anode and cathode materials be electrically connected only through the current-measuring device, a process that involves using an insulating spacer that usually provides enough separation to stop all galvanic action in the atmosphere. There are a few electrochemical techniques that have been applied to atmospheric galvanic corrosion testing, such as the use of a galvanic time-of-wetness meter. Corrosion rates can sometimes be measured indirectly by collecting runoff water and analyzing it for the metals of interest.

Long-Exposure Duration

Atmospheric corrosion, even when accelerated by galvanic effects, is usually a very slow process. Because sensitive electrochemical techniques are difficult to apply in the atmosphere, other less-sensitive techniques must be employed, which means longer exposure periods to make sure that sufficient attack occurs for results to be accurately measurable. The shortest valid test is usually three months, and more typically, exposure periods of one to several years are used. In rare instances, even 30- to 50-year exposures have been performed. The long exposure durations are also necessary to allow for averaging out such variables as rainfall, temperature, and seasons so that prediction of long-term service performance can be made from the test.

Orientation

Unlike immersion exposures, atmospheric exposures must take into account specimen orientation relative to the sun, prevailing winds, and nearest source of corrodent (such as the sea). Height above ground, shading, and even possible abrasion from blown sand in beach exposures must be considered. Oftentimes it is appropriate to follow the guidelines set forth by ASTM Standard G50 on "Conducting Atmospheric Corrosion Tests on Metals," which recommends exposure at an angle of 30 degrees facing south (or the ocean if it is close) and at least 760 mm above the ground.[1]

Wire-on-Bolt (CLIMAT) Test

Advantages/Disadvantages

The wire-on-bolt test was first described by Compton, et al. at Bell Labs in 1955 and has since been used extensively by Alcan under the name CLIMAT Test

(CLassify Industrial and Marine ATmospheres) to determine corrosiveness of atmospheres. The test was recently standardized by ASTM as Standard G116.[2] It consists of wrapping a wire of the anode material around the threads of a bolt or threaded rod of the cathode material and determining mass loss of the anode wire after exposure.

The advantages and disadvantages of this test are summarized in Table 5.1. The small size of the wire, even when compared to the short galvanic interaction distance in atmospheric exposures, gives a large cathode-to-anode area ratio, which accelerates the galvanic attack. The area between the wire and the threads creates a long, tight crevice, also accelerating the corrosion. For these reasons, the CLIMAT test is the most rapid atmospheric galvanic corrosion test, with a typical exposure period of 90 days. It is also the most reproducible of the atmospheric galvanic corrosion tests.

Table 5.1
Wire-on-Bolt (CLIMAT) Test

Advantages	Disadvantages
Fast	Anode and cathode materials must be available in wire and threaded rod forms, respectively
Reproducible	Galvanic polarity must be known
Allows evaluation of seasonal atmospheric variability	Hard to see corrosion form because of size and shape of assembly
	Cannot assess mechanical property decreases
	Staggered exposures must be used to average seasonal effects

The major disadvantages of this test are that the materials must be available in the appropriate wire or threaded rod forms; the more anodic material of the pair must be known beforehand; and the form of the corrosion attack, or its effect on mechanical properties of the base materials, cannot be assessed easily. The short test duration requires the use of staggered tests if seasonal atmospheric variations are to be averaged out to determine in-service performance.

Setup

The components used to construct the specimen assemblies for this test are shown in Figure 5.1. A number of 1-m lengths of 0.875-mm diameter wire of the anode material are prepared. The cathode material is made into 1/2-UNC threaded rods or bolts, which are 12.6 mm in diameter and 100-mm long with approximately 0.5 threads per mm (13 threads per inch). Similar rods of a nonconductive material such as a polyamide (Nylon 66) are also prepared; they will form the same type of crevice as the cathode rods but without the galvanic effect.

All parts should be thoroughly cleaned and degreased before assembly according to the guidelines of ASTM Standard G1 on "Preparing, Cleaning, and Evaluating Corrosion Test Specimens."[3]

A specimen assembly is made by attaching one end of a preweighed wire to a spot near the end of a threaded rod by means of small nuts and bolts. The wire is then wound tightly around one end of the rod by hand or using using a special jig so that the wire lies inside the threads. Figure 5.2 shows the jig used for keeping a constant tension on the wire while it is being wound. While winding, the experimenter should wear clean cotton gloves to prevent contamination of the surfaces of the wire or rod.

The wire should be in contact with a length of threaded rod of roughly 50 mm. The other wire end is then secured to the rod with another set of small nuts and bolts. This hardware should be made of nylon, acetyl resin, TFE-fluorocarbon, or stainless steel insulated with nylon, although use of the rod material is also acceptable. Plastic washers are usually used under the heads of the bolts. In some cases, the wire may be secured to the bolt by means of tight O-rings wrapped around the rod and wire together. After assembly, the excess wire is clipped off and weighed to determine accurately the starting mass of the remaining wire. The assemblies are held vertical by screwing the rod ends farthest from the wire into plastic plates. Figure 5.3 shows a schematic of a completed assembly, and Figure 5.4 is a photograph of an actual completed assembly just before exposure (see page 54). The assemblies are exposed typically for 90 days in the atmospheric site of interest.

Measurements and Interpretation

After exposure, the specimens are visually inspected and the condition of the wires noted (amount of corro-

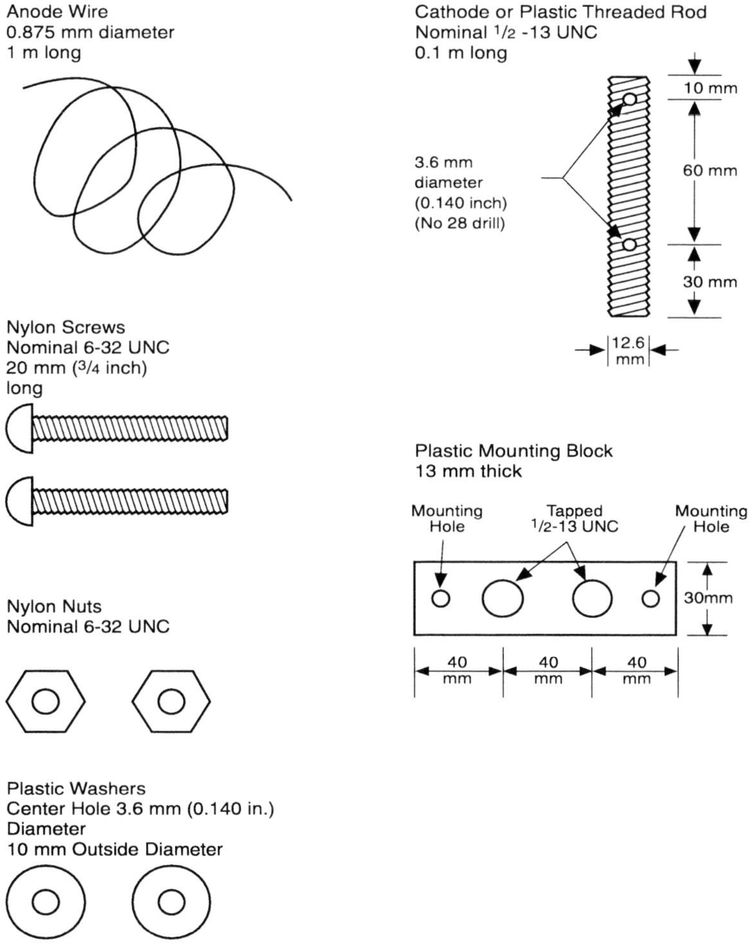

FIGURE 5.1 - Components for making wire-on-bolt exposure assemblies (Courtesy of ASTM).[2]

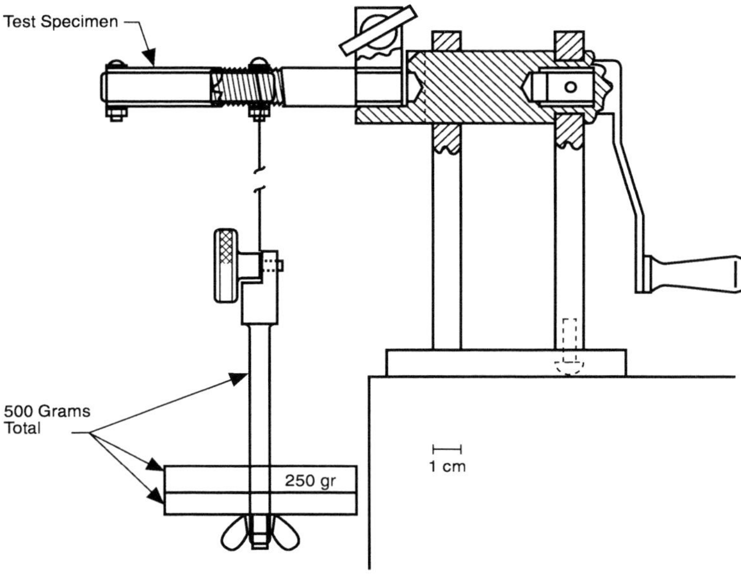

FIGURE 5.2 - Constant tension coil winder for wrapping wire on threaded rods (Courtesy of Alcan International Limited).[4]

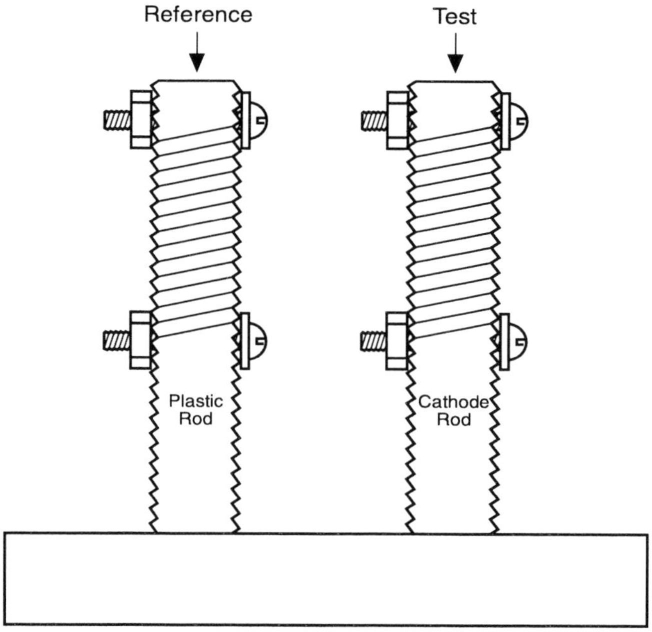

FIGURE 5.3 - Schematic of completed wire-on-bolt assemblies mounted for exposure (Courtesy of Alcan International Limited).[4]

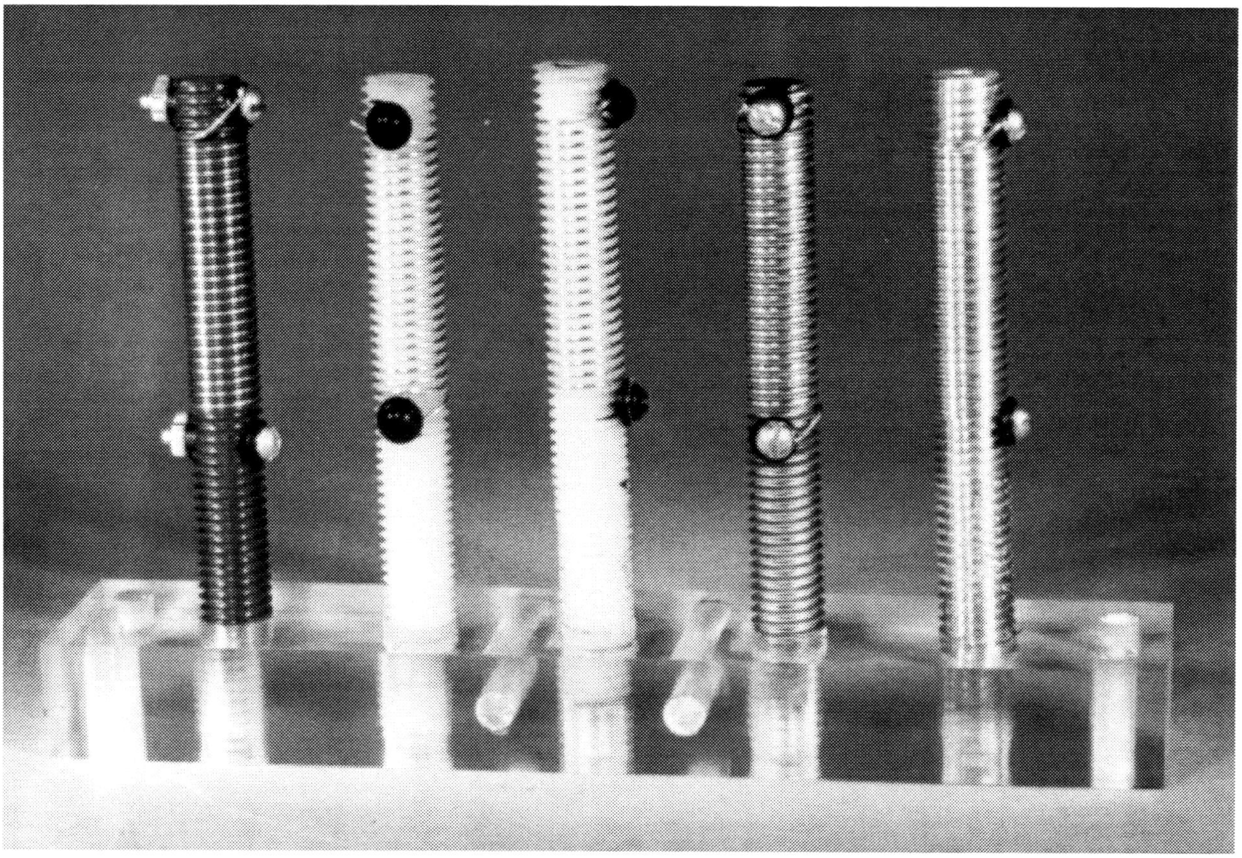

FIGURE 5.4 - Photograph of a completed assembly just before exposure (Courtesy of Alcan International Limited).[4]

sion products, wire thinning, pitting, cracking, and so on). Next, the wires are removed and cleaned according to the procedures specified in ASTM Standard G1 for the material involved.[3] The wires are then weighed and the mass loss determined. The wires exposed on plastic rods are used for reference because they will have experienced no galvanic effects, while the test wires on the cathode rods will have experienced the same corrosion as the references on the plastic rods plus the galvanic action. It is the difference between the mass loss of the wires on the cathode rods and the mass loss of those on the plastic rods that indicates galvanic corrosion.

Because the length of wire actually exposed will be slightly different for each assembly, the length differences may be corrected for if the mass is known for both the original wire and the sections clipped off. The mass loss of the wire may then be corrected to that for a "standard" 1-m length as follows:

$$M_i = M_o - M_E$$

$$M_L = M_i - M_F$$

$$\text{mass loss per meter} = M_L \times \frac{M_o}{M_i}$$

where: M_i = initial mass

M_o = original wire mass

M_E = excess wire mass removal

M_L = mass loss

M_F = final mass (after exposure)

The mass loss must then be normalized to a 90-day exposure period by dividing by the actual number of days of exposure and multiplying by 90. Galvanic effects are calculated as the percent difference in the mass loss between wires in the test and wires in the reference assemblies as follows:

$$\text{galvanic effect} = \frac{\dfrac{\text{test mass loss}}{\text{meter}} - \dfrac{\text{reference mass loss}}{\text{meter}}}{\dfrac{\text{reference mass loss}}{\text{meter}}} \times 100\%$$

For example:

	Reference Wire	Test Wire
M_o	28.1 g	28.4 g
M_E	2.5 g	2.9 g
M_i	28.1-2.5=25.6 g	28.4-2.9=25.5 g
M_F	22.4 g	18.9 g
M_L	25.6-22.4=3.2 g	25.5-18.9=6.6 g
*	$3.2 \times \frac{28.1}{25.6} = 3.51$ g/m	$6.6 \times \frac{28.4}{25.5} = 7.35$ g/m

*mass loss per meter

galvanic effect (7.35-3.51)/3.51x 100 = 109%

If it is found after exposure that the wire on the cathode rod lost less mass than the reference, then it is likely that the wrong material was assumed to be the anode at the outset, and another exposure must be conducted with the roles of the two metallic materials reversed. If the galvanic relationship between the two materials is in doubt and time is limited, exposures should be conducted for both anode/cathode possibilities. Depending on the material combinations selected, longer or shorter exposure durations may be needed to get measurable mass loss or to prevent loss of the wire during exposure. The manufacturing process used to make the wire and rod may affect their corrosion potential. Although unlikely, this may lead to a situation where reversing the materials may also reverse their relationship, making it impossible to construct the apparatus with an anodic wire.

Plate Test

Advantages/Disadvantages

The plate test has been used by a number of experimenters over the years and was first standardized by the International Organization for Standards as ISO 7441.[5] A version has also been standardized by ASTM as Standard G104.[6] The test employs a large plate of the anode material and smaller plates of the cathode material bolted together. The attack on the anode material occurs close to the cathode plates.

The advantages and disadvantages of this test are summarized in Table 5.2 (see Page 54). The arrangement of the plates combined with the short galvanic interaction distance in atmospheric exposures gives a 1:1 cathode-to-anode area ratio, making this test less severe than the CLIMAT test described above. Typical exposure periods range from 1 to 20 years. The area between the plates creates a deep crevice that can hold water for some time and concentrate corrodents. In some cases, the crevice area is coated with an organic such as epoxy before exposure to minimize crevice effects.

A possible disadvantage of this test is that the mate-

Table 5.2 — Plate Test

Advantages	Disadvantages
Can assess corrosion morphology	Slower
Can assess mechanical property decreases	Need sheet material
Standardized	Must know which material is anode

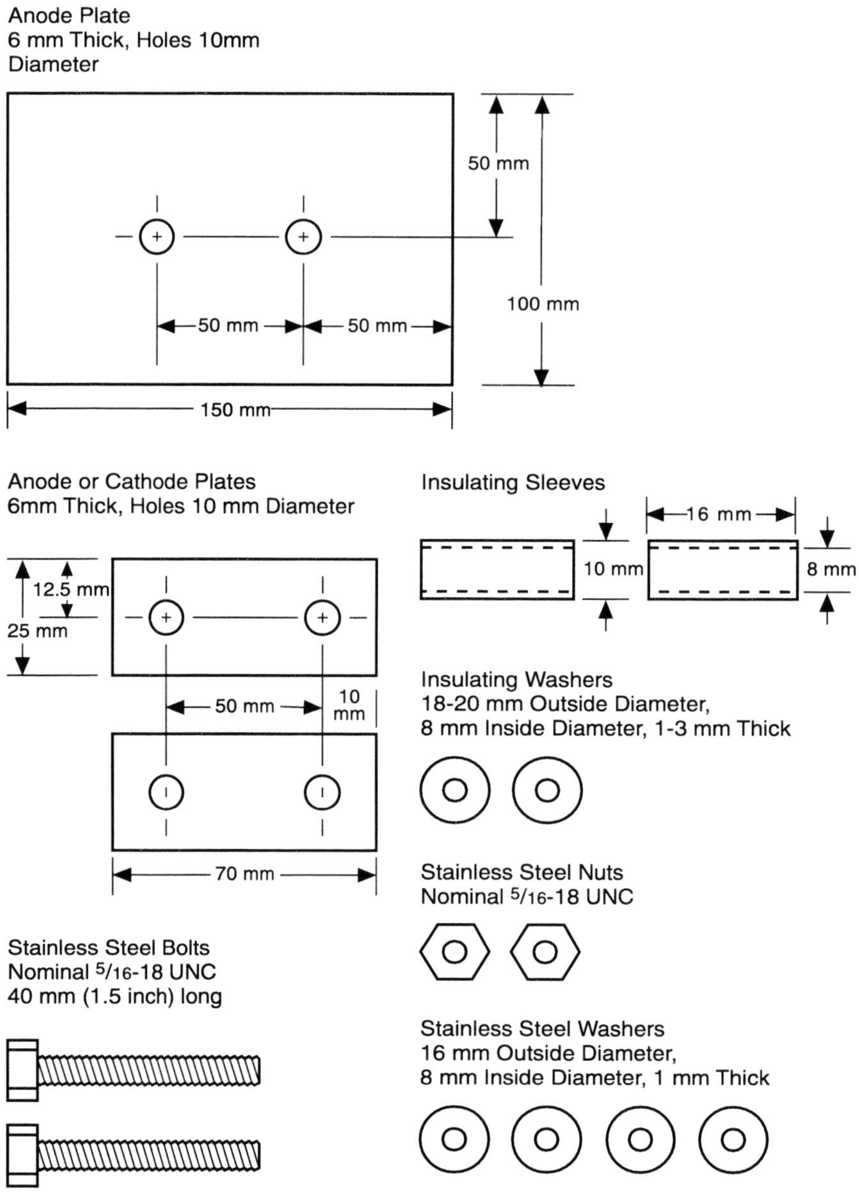

FIGURE 5.5 - Components for making plate exposure assemblies.

rials must be available in the appropriate sheet or plate forms. Another disadvantage is that the anode material of the pair must be known beforehand. This is necessary as the galvanic corrosion must occur on the larger plate, from which mechanical property and corrosion morphology data will be obtained. Although this test takes longer than the CLIMAT test, it has a significant advantage in that corrosion morphology and mechanical properties can be determined after exposure, as well as the mass loss of the anode plate, giving a direct indication of the deterioration of material properties caused by galvanic corrosion.

Setup

Preparation of Plates

Drawings of the components of the specimen assemblies for this test are shown in Figure 5.5. The anode plate is 100 mm by 150 mm and 6 mm thick, with two holes 50 mm apart for the bolts that hold on the cathode plates. If mechanical testing is not to be done on the anode plates, alternative plate sizes with the same bolt holes can be used instead. Other sizes suggested in the ISO standard are 150 mm by 90 mm, 240 mm by 90 mm, and 65 mm by 110 mm. The cathode plates are always 25 mm by 70 mm and 6 mm thick, with two holes for the attachment bolts.

Location of the holes and relative orientation of the plates are shown in Figure 5.5. The thickness of the cathode plates is not critical, although if they are much thinner than 6 mm, they will not be stiff enough to maintain even tightness in the center area between the bolts. The anode plate's thickness is, however, important because equivalent depths of attack will have a greater effect on the strength of a thin plate than on the strength of a thick one. If any plate's thickness is significantly different from 6 mm, that will require that the insulating sleeve lengths in Figure 5.5 be adjusted accordingly.

As with the CLIMAT test, if the relative anode/cathode relationship of the two materials is not known beforehand, extra assemblies will have to be tested with the anode and cathode materials reversed. The edges of the specimens and bolt holes must be free of metal burrs, and the surfaces should be free from visible defects such as heat-treatment scale, blisters, cracks, porosity, scratches, and so on. A surface finish of 0.4 to 0.5 micrometer is usually suggested. The specimens may be marked in areas well away from the contact area by any method desired. Three methods are typical: punching of numbers, cutting of edge notches, or attachment of metal tags through holes drilled near the edge of the anode plate. Metal tags may be made from any material that will not corrode much over the test period, such as 70-30 nickel-copper, titanium alloys, austenitic stainless steels, nickel alloy 625, aluminum, or copper-nickel. Galvanic corrosion of the plate caused by the tag is not important for mechanical property determination although if mass loss is to be determined, a galvanically compatible tag material should be used. In no case should copper-bearing tag material be used on aluminum plates or aluminum tags be used on copper-bearing plates because copper corrosion products can accelerate the corrosion of aluminum alloys.

Holes in the plates should be approximately 10 mm in diameter and should be drilled, not punched, to prevent excessive deformation at the edges. The bolts used to secure the specimens are typically 8 mm by 40 mm and should be made of a material resistant to the atmosphere for the duration of the exposure. Monel Alloy 400, titanium alloys, and some stainless steels such as type 316 or 304 are usually sufficient for most atmospheres. Construction of a plate assembly is illustrated in Figure 5.6. The bolts are used with nonmetallic insulating washers and sleeves, as shown in Figure 5.6, in order to

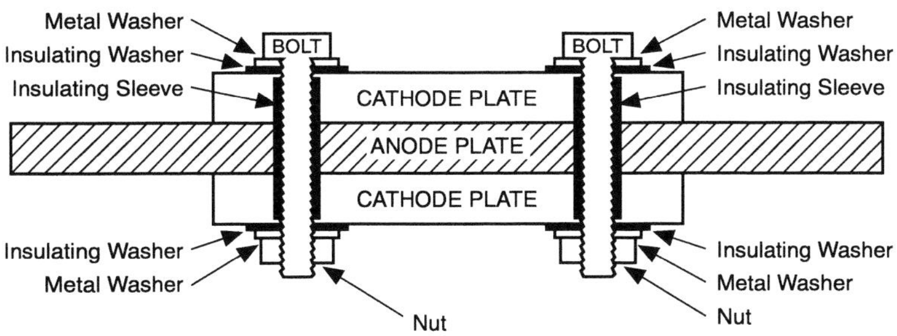

FIGURE 5.6 - Construction of plate assembly.

prevent electrical contact between the bolts and the plates.

Additional identical plates are usually prepared for control specimens that will be stored in airtight containers during the exposure to assess the original mechanical properties and effects of post-test cleaning and other conditions not related to the atmospheric exposure. A set of plates for reference assemblies, where both the large and the small plates are made of the anode material, is also prepared for exposure under identical conditions to those seen by the test assemblies. The reference assemblies are used to assess nongalvanic effects such as general or crevice corrosion, or pitting.

It is usually recommended that, for optimum statistical validity, at least three assemblies of each type be prepared and tested for each exposure duration to be evaluated.

Assembly and Exposure

Immediately before exposure, the specimens should be cleaned following procedures in ASTM Standard G1 and degreased with appropriate organic solvents such as chlorinated solvents (trichloroethane), alcohols, or petroleum spirits (benzene); and the initial mass of the anode plates should be determined.[3] Many effective degreasing solvents are carcinogenic and flammable, so appropriate precautions should be taken (a good fume hood, no sparks, safety glasses, gloves, etc.). After the plates have been degreased, the experimenter should handle them only by the edges and only while wearing clean cotton gloves. In cases where crevice corrosion is expected to be a large problem or where corrosion products may force the plates apart and break the electrical connection between them, a thin layer of an organic coating may be applied to both sides of the anode plate so that the coating completely covers the area between the anode and cathode plates without projecting beyond the cathode plates. The dry coating thickness should not exceed 10 micrometers and the coating should not be applied to areas within approximately 2 to 5 mm of the bolt hole edges so that electrical contact is achieved between the plates after assembly.

In some cases, use of a coating may not prevent the anticipated problems, and it may complicate maintenance of electrical contact between the anode and cathode plates. Use of a coating also complicates area calculations for results expressed on a per-unit-area basis. If coating deterioration occurs, the length of exposure of areas under the coating that are subsequently exposed is not well defined. For these reasons many experimenters do not use coatings.

Tightening torque for the bolts is not given in the ASTM or ISO standards on this test method. Overtightening will lead to cracking of the insulators, causing the bolts to short to the plates. Undertightening will result in electrical resistance between the plates that is too high. Electrical resistance of less than 1 ohm between each cathode plate and the anode plate should be assured by the use of a resistance meter. Electrical insulation of each bolt from the three plates in each assembly should also be ensured, with the measured resistance no lower than 10,000 ohms. These electrical resistance checks should be repeated on the dry assemblies occasionally throughout the specimen exposure period. If resistance between the plates rises above 1 ohm during the exposure, the bolts should be retightened without disturbing the assemblies. If resistance between the bolts and the plates falls below 10,000 ohms, check for conductive surface deposits, which could be causing a short circuit, and carefully remove them. If the reason for the lower resistance cannot be found and corrected without disturbing the plates, the test may be invalid unless the bolt material is galvanically the same as one of the plate materials.

Throughout the duration of the exposure, the control specimens are stored in airtight containers with less than 50 percent relative humidity, for example, in a desiccator or a plastic bag containing silica gel or activated alumina. Some plastics may give off fumes that are corrosive to certain metals, so the type of plastic bag should be selected with caution and after consultation with a corrosion engineer.

Test and reference assemblies should be mounted on racks at the ISO standard angle of 30 or 45 degrees from the horizontal (30 degrees is the historically preferred angle in North America, 45 degrees in Europe) with the short axis of the anode plates parallel to the ground. The racks should meet the specifications of ASTM Standard G50 regarding construction and positioning (facing south or towards the ocean if it is nearby), and the specimens should be mounted using insulators to prevent electrical contact with the metal racks.[1] A rack-mounted assembly is shown in Figure 5.7. Typical exposure periods range from one to 20 years, with removals spaced at logarithmic intervals such as 1, 2, 5, 10, and 20 years.

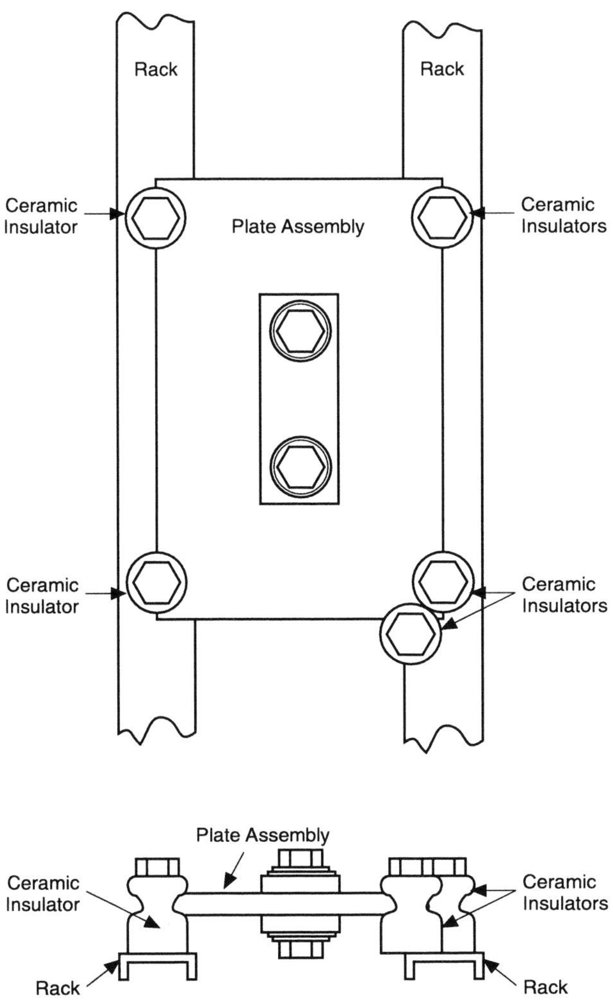

FIGURE 5.7 - Rack-mounted plate assembly (Courtesy of ASTM).[6]

Measurements and Interpretation

After the assemblies are removed from the testing site, their surface appearance should be noted. In particular, comparisons of both the upper and lower assembly surfaces of the control, reference, and test assemblies should be made. Visual examination should be made for tarnishing, the amount and color of corrosion products, and the character and distribution of attack (uniform corrosion, pitting, crevice corrosion between plates, attack at edges, near cathode plate, etc.). Next, the assemblies are taken apart and any additional observations noted, such as the condition of the coating (if used) or of the surfaces under the cathode plates.

The anode plates can now be cleaned following the procedures in ASTM Standard G1, rinsed, dried, and stored in a desiccator for 24 hours (to ensure complete drying) prior to weighing. Drying may also be accomplished by the use of a low-temperature oven. Final mass can next be determined for mass-loss calculations and, if desired, corrosion rate can be calculated based on the total uncoated area and the procedures in ASTM Standard G1.[3]

If a coating has been used and has deteriorated, corrosion rate calculation is not recommended because of uncertainty about the duration of exposure of the areas under the degraded coating. If necessary, a pitting evaluation can be conducted according to ASTM Standard G46, which involves measuring the average number of pits per unit area and the pit depths.[7] Depth of any localized attack, such as that near the cathode plates, should next be measured using depth gauges, roughness profile meters, or metallographic techniques. If desired, metallographic samples can be taken from the locations on the plates shown in Figure 5.8, which are representative of crevice and noncrevice areas next to the cathode plates. Metallography can be used for determining if the corrosion occurred preferentially between the metal grains, if delamination due to corrosion product wedging (exfoliation) has occurred, and so on.

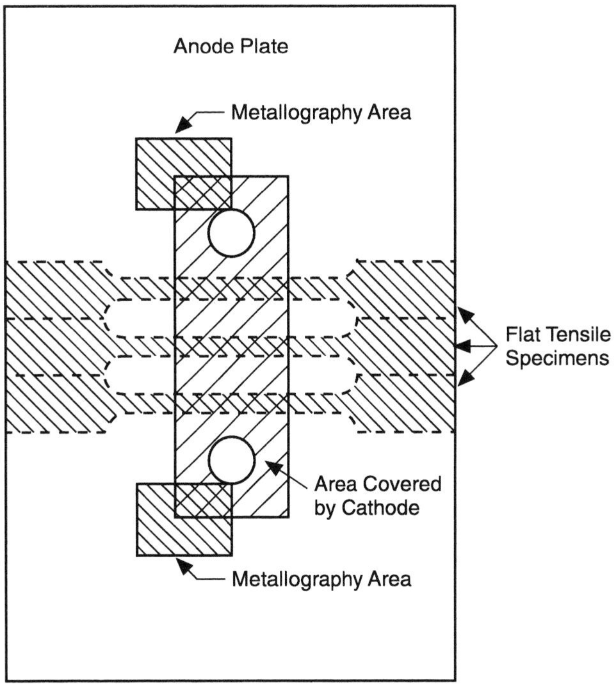

FIGURE 5.8 - Location of areas in anode plate for post-exposure testing.

If mechanical property information is desired, the anode plates are cut into flat tensile specimens, machined to size according to Figure 5.8 and the procedures of ASTM Standard Test Method E8, and tested according to the Standard.[8] Typical mechanical properties that can be determined are: breaking load, ultimate tensile strength, yield strength, and elongation. Those properties that rely on a measure of cross-sectional area after corrosion may be difficult to determine accurately because nonuniformity of corrosion attack makes cross-sectional area measurements difficult.

Nongalvanic corrosion effects of the exposure can be determined by calculating the percent difference in the property of interest between reference and control specimens as follows:

$$\text{non-galvanic effect(\%)} = \frac{\text{ref. property-control property}}{\text{control property}} \times 100$$

Galvanic effects can be separated out by calculating the percent difference in the property of interest between test and reference specimens as follows:

$$\text{galvanic effect(\%)} = \frac{\text{test property - reference property}}{\text{reference property}} \times 100$$

The total effect of the exposure can be determined by calculating the percent difference in the property of interest between the test and control specimens as follows:

$$\text{total exposure effect(\%)} = \frac{\text{test property - control property}}{\text{control property}} \times 100$$

Washer Test

Advantages/Disadvantages

The washer, or disk, test was used by H.O. Teeple in 1949 for a series of exposures for ASTM Committee B3.08 on Corrosion of Non-Ferrous Metals and Alloys. Since that time, ASTM has refined the test and conducted other investigations using this method, and an ASTM standard for this method is under development. A specimen assembly consists of a set of four washers, two of the anodic material and two of the cathodic. These alternate in the test assembly. The washers are graduated in size and bolted together through center holes. Only the center two washers are used for the analysis; the outer washers are used only to establish the proper environment on the inner washers. The advantages and disadvantages of the washer test are summarized in Table 5.3.

Table 5.3
Washer Test

Advantages	Disadvantages
Easily obtained thin sheet can be used	Slower
Easy	Hard to see details of corrosion morphology
Galvanic relationship need not be known	Cannot assess mechanical property decreases

The arrangement of the washers gives close to a 1:1 cathode-to-anode area ratio, making this test less severe than the CLIMAT test described above and similar in severity to the plate test. As in the plate test, typical exposure periods range from 1 to 20 years. Of the tests described in this chapter, this is probably the easiest one for which to obtain materials, as only small pieces of thin sheet are required. The anode/cathode relationship of the materials involved need not be known beforehand.

Unlike the plate test, the washer test yields only corrosion rate information, and it is difficult to determine corrosion morphology visually because of the shape of the components. It is also difficult to determine changes in mechanical properties with this test. Of the three tests discussed in this chapter, the washer test is probably the easiest to set up and evaluate.

Setup

Preparation of Washers

Drawings of the components needed to manufacture specimen assemblies for this test are shown in Figure

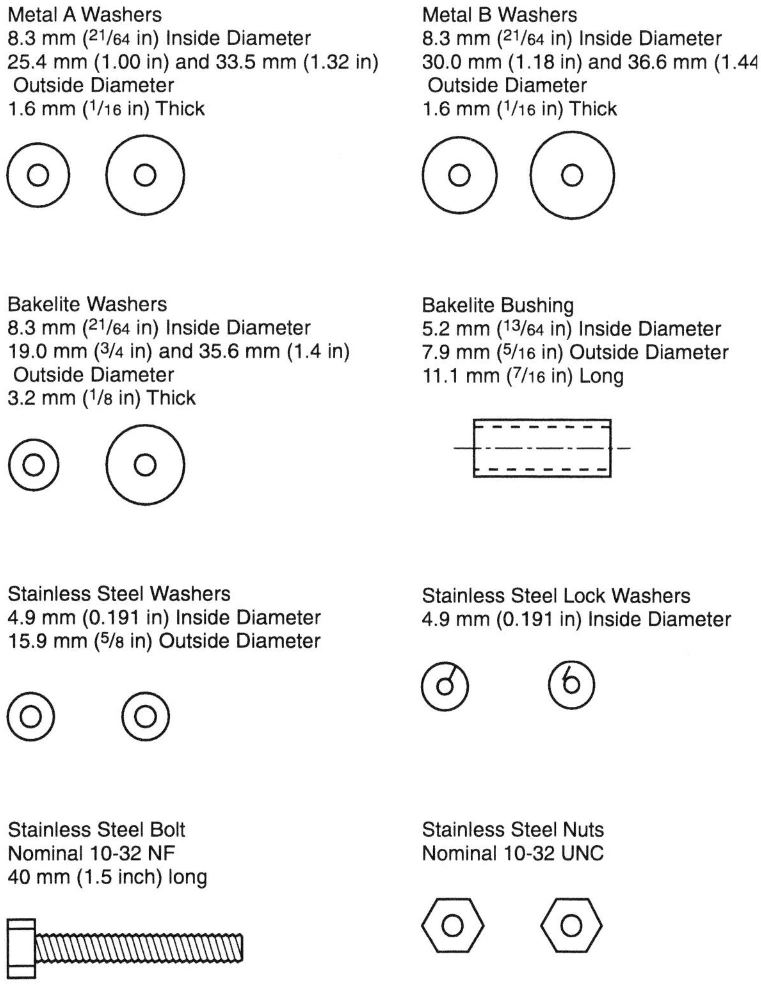

FIGURE 5.9 - Components for making washer exposure assemblies.

5.9. Washers of one material are manufactured 1.6 mm thick with outside diameters of 25.4 mm and 33.5 mm. The other material in the galvanic couple is made into washers of identical thickness with outside diameters of 30.0 mm and 36.6 mm. Washer dimensions are critical to maintaining the 1:1 area ratio fixed by the geometry. Thickness should be adjusted by surface grinding if required. Surface finish should be 32 RMS or smoother. Bakelite† washers 3.2-mm thick by 19.0 mm and 35.6 mm in outside diameter are also manufactured. Also used in the assembly is a Bakelite bushing 5.2 mm in inside diameter, 7.9 mm in outside diameter, and 11.1 mm long. Other materials besides Bakelite may be used provided they are nonconductive, hard, and do not retain significant moisture. Two stainless steel washers 15.9 mm in outside diameter and a 4.8-mm by 38-mm stainless steel bolt with two nuts and lockwashers complete the list of materials. All washers of all test materials have a central hole just large enough to clear the 7.9-mm diameter bushing. The stainless steel washers should have an inside diameter just sufficient to clear the 4.8-mm diameter bolt. Washers should not be made by punching directly to the final dimensions, as this introduces a deformed area near the edge. Washers may be made by first punching out disks a minimum of 3 mm too large in diameter, machining them to final dimensions, and then drilling holes in the disk centers. This procedure should avoid the problem of deformities near the edge.

† Trade name

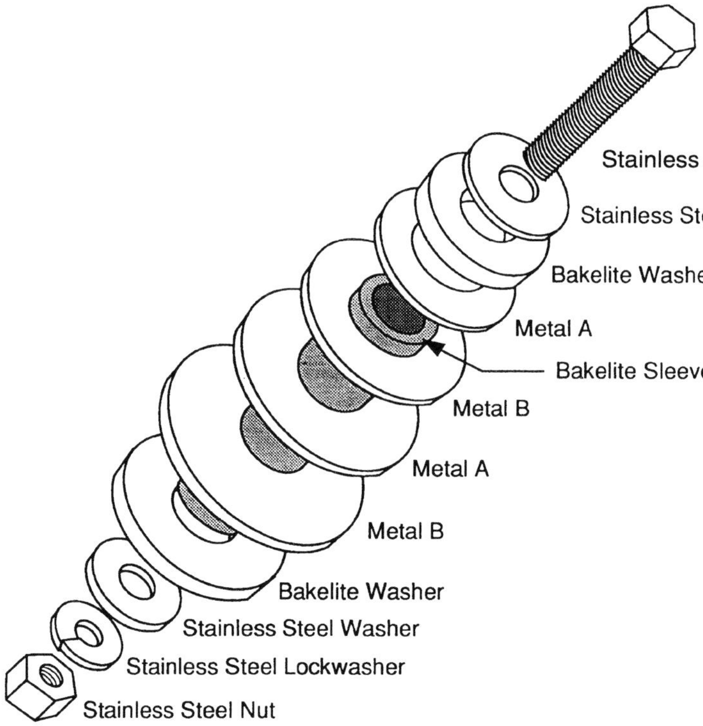

FIGURE 5.10 - Exploded view of washer test assembly.

The edges of the specimens and bolt holes must have metal burrs removed, and the surfaces should be free from visible defects such as heat-treatment scale, blisters, cracks, porosity, scratches, etc. A surface finish of 0.4 to 0.5 µm is usually suggested. The specimens may not be marked because there is no appropriate place to do so; however, metal tags may be affixed to the completed assemblies outside of the Bakelite washers if desired. Although Figures 5.9 to 5.11 indicate the use of stainless steel for the bolts, end washers, and lockwashers, this and other hardware may be made of any material provided that material will not degrade in the atmosphere for the duration of the exposure. Monel Alloy 400 and titanium alloys are usually sufficient for more aggressive atmospheres, while aluminum or austenitic (18% Cr, 8% Ni) stainless steels may be sufficient for less aggressive atmospheres.

Additional washers of the anode material may be prepared to enable the construction of reference assemblies, where all four washers are made of the anode material and exposed to the same conditions as the test assemblies. Where the anode material is not known beforehand, reference assemblies of both materials should be constructed. The reference assemblies are used to assess nongalvanic effects such as general or crevice corrosion, or exfoliation.

It is usually recommended that, for optimum statistical validity, at least three assemblies of each type be prepared and tested for each exposure duration evaluated.

Assembly and Exposure

Immediately before exposure, the specimens should be cleaned following procedures in ASTM Standard G1 and degreased with appropriate organic solvents such as chlorinated solvents (trichloroethane), alcohols, or petroleum spirits (benzene); and the initial mass of the anode plates should be determined.[3] Many effective degreasing solvents are carcinogenic and flammable, so appropriate precautions should be taken (a good fume hood, no sparks, safety glasses, gloves, etc.). If the anode/cathode relationship of the materials is not known at the start of the test, all washers should be weighed before assembly. After the washers have been degreased, the experimenter should handle them only by the edges and only while wearing clean cotton gloves.

The specimens are next assembled as shown in Fig-

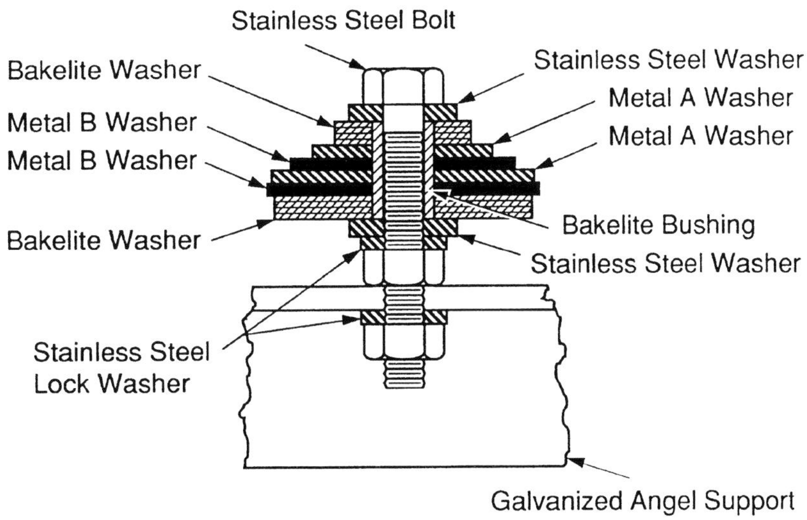

FIGURE 5.11 - Cross section of washer test assembly (Courtesy of ASTM).[9]

ure 5.10. The bolt should be tightened firmly with a wrench until there is good electrical contact between washers. Insufficient torque will cause too high a resistance between the washers, and too much torque will crack the bushing and cause the bolt to short to the washers. Electrical resistance of less than 1 ohm between neighboring washers should be verified by the use of a resistance meter. Electrical insulation of the bolt from the washers should also be verified, with the measured resistance no lower than 10,000 ohms. These electrical resistance checks should be repeated on dry assemblies occasionally throughout the specimen exposure period, and if the resistance increases, the bolts should be retightened without disturbing the assemblies. Both the test and reference assemblies are mounted vertically in holes drilled in metal angle iron and held in place by lockwashers and nuts as shown in Figure 5.11. Typical exposure periods range from one to 20 years, with removals spaced at logarithmic intervals such as 1, 2, 5, 10, and 20 years.

Measurements and Interpretation

After the assemblies are removed from the testing site, their surface appearance should be noted. In particular, comparisons between the control and reference assemblies should be made. Visual examination should be made for tarnishing, the amount and color of corrosion products, and the character of attack (uniform, crevice, edge, and so on). Next, the assemblies are taken apart and any additional observations noted, such as the condition of the contacting surfaces. The washers can now be cleaned following the procedures in ASTM Standard G1, rinsed, dried, and stored in a desiccator for 24 hours (to ensure complete drying) prior to weighing.[3] Final mass can next be determined for mass-loss calculations. If desired, metallographic samples can be taken from the washers to determine if the corrosion is intergranular, if exfoliation has occurred, and so on.

Galvanic effects are determined by calculating the percent difference in the mass loss between anode washers in the test and reference assemblies as follows:

$$\text{galvanic effect(\%)} = \frac{\text{test mass loss-reference mass loss}}{\text{reference mass loss}} \times 100$$

Special Considerations in Atmospheric Testing

There are a number of special considerations in atmospheric testing. The atmosphere can cause unusual test problems. For example, the author had atmospheric specimens exposed at a beach site when a hurricane hit. Although the specimens were not lost, they were cleaned of corrosion products down to bright metal by wind-blown sand. In grass lots, the use of weed or bug killers or defoliants under the specimen racks can create fumes that affect corrosion behavior of the samples. Several works cited in the bibliography can give guidance on some of these more unusual considerations. As in all corrosion testing, the best guides are knowledge of what affects corrosion and use of common sense when setting up the tests.

References

1. ASTM G50 E1-76,"Standard Practice for Preparation and Use of Direct Tension Stress-Corrosion Test Specimens." (Philadelphia, PA: ASTM, 1976).

2. ASTM G116-93, "Practice for Conducting the Wire-on-Bolt Test for Atmospheric, Galvanic Corrosion." (Philadelphia, PA: ASTM, 1993).

3. ASTM G1-90, "Standard Practice for Preparing, Cleaning, and Evaluating Corrosion Test Specimens." (Philadelphia, PA: ASTM, 1990).

4. Alcan International Ltd. Research and Development Center (Kingston, Ontario, Canada: 1993).

5. ISO 7441-84, "Corrosion of Metals and Alloys - Determination of Bimetallic Corrosion in Outdoor Exposure Corrosion Tests." (Geneva, Switzerland: ISO, 1984).

6. ASTM G104-89, "Standard Test Method for Assessing Galvanic Corrosion Caused by the Atmosphere." (Philadelphia, PA: ASTM, 1989).

7. ASTM G46-76, "Standard Practice for Examination and Evaluation of Pitting Corrosion." (Philadelphia, PA: ASTM, 1976).

8. ASTM E8-93, "Standard Test Methods for Tension Testing of Metallic Materials." (Philadelphia, PA: ASTM, 1993).

8. ASTM E8-93, "Standard Test Methods for Tension Testing of Metallic Materials." (Philadelphia, PA: ASTM, 1993).

9. ASTM STP646-78, "Atmospheric Factors Affecting the Corrosion of Engineering Metals." (Philadelphia, PA: ASTM, 1978).

Bibliography

General

Ailor, W., ed. *Handbook on Corrosion Testing and Evaluation.* New York, NY: John Wiley and Sons, Inc., 1963.

Baboian, R., G. Haynes, P. Sexton. "Atmospheric Corrosion of Laminar Composites Consisting of Copper on Stainless Steel." *Atmospheric Factors Affecting the Corrosion of Engineering Materials*, ed. S. Coburn. ASTM STP 646. Philadelphia, PA: ASTM, 1978.

Compton, K.G., A. Mendizza, W.W. Bradley, *Corrosion* 11, 9 (1955): p. 35.

LaQue, F.L. "Corrosion Testing." Edgar Marburg Lecture. Philadelphia, PA: ASTM, 1951, pp. 53-58.

"Report of Subcommittee VIII, of Committee B-3." *Proceedings*, ASTM. Vol. 284, No. 1. Philadelphia, PA: ASTM, 1928.

Wire-on-Bolt Test

Addanki, R.S., K.P. Mukherjee, A.K. Lahiri. "Galvanic Corrosion of Ferrous and Non-Ferrous Metals in the Industrial Atmosphere of Jamshedpur." *National Metallurgical Laboratory Technical Journal* (India), 4, 17 (1962): pp. 17-22.

Arlery, R. *Congreys International de la Corrosion Marine et de Salissures*, held June 1964. Paris: Editions du Centre de Recherches et d'Etudes Oceanographiques, 1965.

Doyle, D.P., H.P. Godard. "A Rapid Method for Determining the Corrosivity of the Atmosphere at Any Location." *Nature* 200, 4912 (1963): pp. 1167-1168.

Doyle, D.P., H.P. Godard. "A Rapid Method for Predicting Adequate Service Lives for Overhead Conductors in Marine Atmospheres." 1971 IEEE Winter Power Meeting, paper no. 71 CP 172-PWR. New York, NY: IEEE, 1971.

Doyle, D.P., H.P. Godard. "Corrosivity of the East London and Port Elizabeth Atmospheres to Aluminum—Influence of Distance from the Ocean." *Corrosion and Metal Finishing* (South Africa) 1, 2 (1964): p. 10.

Doyle, D.P., H.P. Godard. "Corrosivity of the East London and Port Elizabeth Atmospheres to Aluminum—Influence of Distance from the Ocean." *Corrosion and Metal Finishing* (South Africa) 1, 10 (1964): p. 8.

Doyle, D.P., H.P. Godard. "Rapid Determination of Corrosivity of an Atmosphere to Aluminum." *Proceedings of the Third International Congress on Metallic Corrosion*. Vol. 4. Moscow: MIR Publishers, 1969.

Doyle, D.P., T.E. Wright. "Rapid Methods for Determining Atmospheric Corrosivity and Corrosion Resistance." *Atmospheric Corrosion*, ed. W. Ailor. New York, NY: John Wiley and Sons, Inc., 1982.

Godard, H.P. *Materials Protection* 2, 6 (1963): p. 38.

Hache, A. *Corrosion et. Anti-corrosion* 10, 3 (1962): pp. 68-77.

Washer Test

Baboian, R. "Final Report on the ASTM Study: Atmospheric Galvanic Corrosion of Magnesium Coupled to Other Metals." *Atmospheric Factors Affecting the Corrosion of Engineering Materials*, ed. S. Coburn. ASTM STP 646. Philadelphia, PA: ASTM, 1978.

Blum, W. *Symposium on Properties, Tests and Performance of Electrodeposited Metallic Coatings*. ASTM STP 197. Philadelphia, PA: ASTM, 1956.

Proceedings, ASTM. Vol. 41, No. 170. Philadelphia, PA: ASTM, 1941.

Teeple, H.O. Symposium on Atmospheric Corrosion of Non-Ferrous Materials. ASTM STP 175. Philadelphia, PA: ASTM, 1955, pp. 89-113.

Index

A

B

C

NOTES